T0275919

POETRY REALIZED IN NATURE

Poetry realized in nature

Samuel Taylor Coleridge and early nineteenth-century science

TREVOR H. LEVERE

Professor of the History of Science
University of Toronto

CAMBRIDGE UNIVERSITY PRESS
CAMBRIDGE
LONDON NEW YORK NEW ROCHELLE
MELBOURNE SYDNEY

PUBLISHED BY THE PRESS SYNDICATE OF THE UNIVERSITY OF CAMBRIDGE
The Pitt Building, Trumpington Street, Cambridge, United Kingdom

CAMBRIDGE UNIVERSITY PRESS
The Edinburgh Building, Cambridge CB2 2RU, UK
40 West 20th Street, New York NY 10011–4211, USA
477 Williamstown Road, Port Melbourne, VIC 3207, Australia
Ruiz de Alarcón 13, 28014 Madrid, Spain
Dock House, The Waterfront, Cape Town 8001, South Africa

http://www.cambridge.org

First published 1981
First paperback edition 2002

A catalogue record for this book is available from the British Library

Library of Congress Cataloguing in Publication data
Levere, Trevor Harvey.
Poetry realized in nature.
Includes index.
1. Coleridge, Samuel Taylor, 1772–1834 – Knowledge – Science.
2. Literature and science. 3. Science –
Great Britain – History – 19th century. I. Title.
PR4487.S35L4 821′.7 81-1930 AACR2

ISBN 0 521 23920 6 hardback
ISBN 0 521 52490 3 paperback

TO KEVIN AND REBECCA

CONTENTS

CHAPTER 3

Two Visions of the World: Coleridge, Natural Philosophy, and the Philosophy of Nature

CHAPTER 4

Coleridge and Metascience: Approaches to Nature and Schemes of the Sciences

CHAPTER 5

The Construction of the World: Genesis, Cosmology, and General Physics

CHAPTER 6
Geology and Chemistry: The Inward Powers of Matter

CHAPTER 7
Life: Crown and Culmination

PREFACE

Several years ago, confronted by problems in the history of chemistry, I sought an interview with Kathleen Coburn, to ask for help in unraveling Humphry Davy's possible indebtedness to Coleridge. Miss Coburn's response was to introduce me to Coleridge's remaining unpublished notebooks, some of them replete with chemical and other scientific entries. She cheerfully invited me to make sense of them. The notes proved, for the most part, individually incomprehensible, forming part of an intellectual enterprise that was then unfamiliar to me. This book has grown from my attempts to overcome that unfamiliarity; and so I take pleasure in reminding Kathleen Coburn that she provided my incentive.

Miss Coburn and Merton Christensen, coeditors of the forthcoming volumes of *The Notebooks of Samuel Taylor Coleridge*, have been generous with criticism, encouragement, information, transcripts, and time. Miss Coburn has also been of great assistance in persuading me of the value of a blue pencil in revising drafts of my manuscript.

Several editors of individual works in *The Collected Works of Samuel Taylor Coleridge* (general editor Kathleen Coburn) have been equally generous. Heather Jackson told me of several manuscripts of which I was ignorant; she allowed me to read drafts of her edition of the *Essay on Scrofula* and the *Theory of Life* for the *Shorter Works and Fragments*; and she read and criticized a complete draft of this book. George Whalley gave me access to the typescript of his monumental edition of the *Marginalia*, of which the first volume (A–B) has been published this year; and he provided advice and encouragement along the way. James Engell and Walter Jackson Bate gave permission for me to examine the typescript of their edition of the *Biographia Literaria*. Barbara Rooke's edition of *The Friend* has time and again proved invaluable for its insights.

All who write on Coleridge must be indebted to Earl Leslie Griggs for his edition of the *Collected Letters of Samuel Taylor Coleridge*, and to Kathleen Coburn for the three volumes of the *Notebooks* that have already appeared.

Owen Barfield's *What Coleridge Thought* and Thomas McFarland's *Coleridge and the Pantheist Tradition* have been the books most in my mind as I pursued Coleridge. L. Pearce Williams first made me think of Coleridge in the context of the history of science, and Craig Miller suggested initial directions and has been constantly encouraging.

My debts are personal as well as intellectual: to my mother and to the memory of my father, who saw education as the key to freedom; to Jennifer, my wife, who provided certainty and support as my ideas took shape; and to my children, who have helped more than they know, and to whom I dedicate this book.

I wish to thank Mrs. Freda Gough for much assistance, notably in matters concerning the Coleridge Collection in Victoria College Library, University of Toronto. I am also grateful to the staff of the British Museum Library and Department of Manuscripts; the Bodleian Library; the Bristol Central Library; the Bristol Record Office; Cambridge University Library; the Wordsworth Collection in the Cornell University Library; the Cornwall Record Office; the Dove Cottage Library; Edinburgh University Library; the Houghton Library of Harvard University; the Henry H. Huntington Library and Art Gallery; the library of Jesus College, Cambridge; Keele University Library, and Dr. Ian Fraser; the Osler Library of McGill University; the New York Academy of Medicine; the Berg Collection in the New York Public Library; the J. Pierpont Morgan Library; the Royal College of Surgeons; the Royal Institution of Great Britain; the Royal Society of London; the Swedenborg Society, London; the Thomas Fisher Library in the University of Toronto; Trinity College, Cambridge; University College Library in the University of London; the library of the Wellcome Institute for the History of Medicine; Dr. Williams's library, London; and the Beinecke Rare Book and Manuscript Library in Yale University, and Miss Marjorie Wynne.

Philip Enros, Kathleen Ochs, and John Wojtowicz provided accurate and imaginative research assistance. Bev Jahnke and Gladys Bacon turned disorderly manuscript into typescript ready for the publisher.

The bulk of the research was carried out during sabbatical leave from the University of Toronto; I am grateful to my colleagues in the Institute for the History and Philosophy of Science and Technology, and to my university, for this privilege. My work has been generously supported by grants from the Canada Council and the Humanities and Social Sciences Research Committee of the University of Toronto, and by a Killam Senior Research Scholarship.

Princeton University Press and Routledge and Kegan Paul, Ltd., have given permission for passages to be quoted from *The Collected Works of Samuel Taylor Coleridge*, ed. K. Coburn, Bollingen Series 75. Passages from the *Notebooks of Samuel Taylor Coleridge* and from the *Collected Letters of Samuel Taylor Coleridge* are quoted by permission of Princeton University Press and Oxford University Press, respectively. I am also grateful to the British Museum Library and Department of Manuscripts for permission to quote from manuscripts in their collections.

I have worked either from published editions, or from original manuscripts or photographs thereof. ⟨Angle brackets⟩ in quotations indicate the original author's insertions in a manuscript. [Square brackets] are used to indicate my own emendations or insertions in a quotation; with the addition of a question mark, square brackets also indicate doubtful readings. Words crossed out in a quotation are those crossed out in the manuscript. I have interpreted Coleridge's editing of the manuscript sources so that his double underlining of capital and lower case letters is set here as a capital and small capitals and his single underlining is set as italics. Works either are cited in full in the text or notes or are cited in abbreviated titles, a list of which precedes the notes.

T.H.L.

Toronto, Ontario
December 1980

INTRODUCTION

NATURE AND MIND

Samuel Taylor Coleridge, in the organic unity of his thought, worked constantly toward a system reducing "all knowledges into harmony."[1] Thomas McFarland has observed that "the urge to system is a reflection, in the special realm of philosophy, of a universal concern, the need to harmonize, to tie things together – what we may call the need for reticulation."[2] This need impelled Coleridge to attempt to reconcile conflicting systems of thought, and to make room for all facts of experience.

He was determined to construct a scheme that was truly comprehensive, encompassing the reality and dignity of external nature, the moral sense and freedom of will of mankind, and God, in whom man had his being.[3] Faith in this scheme was not irrational, but rather subsisted "in the *synthesis* of the reason and the individual will."[4] Trinitarian Christianity came to provide a unifying logic for this scheme, and that logic in turn supported Christianity, in a mutual interdependence: "True philosophy rather leads to Christianity, than contained anything preclusive of it."[5] Coleridge asserted, indeed, "that it was one of the great purposes of Christianity . . . to rouse and emancipate the soul from . . . debasing slavery to the outward senses, to awaken the mind to the true *criteria* of reality, namely, permanence, power, will manifested in act, and truth operating as life."[6]

Coleridge held fast to the reality of man's moral being and to that of external nature. He would not evade, nor, as McFarland has shown, could he fully resolve this tension. Christianity, trinitarian Christianity, enabled him to retain both poles, the real and the ideal: "That which gives a reality to the idea, that which gives the dignity of the ideal to reality, that which combines all the common sense of the experimental philosopher with all the greatest prospects of the Platonists – that we find in Christianity." The Trinity was the central unifying idea, "the primary Idea, out of which all other Ideas are evolved."[7]

Plato, poet and philosopher, interfusing thought and feeling, was one whom Coleridge admired and used in working toward the system that would synthesize religion with philosophy, viewing him as one wholly aware of the need for reticulation: "Plato . . . perceived . . . that the knowledge of man by himself was not practicable without the knowledge of other things, or rather that man was that being in whom it pleased God that the consciousness of others' existence should abide, and that therefore without natural philosophy and without the sciences which led to the knowledge of objects without us, man himself would not be man."[8] These were Coleridge's own convictions. Poet, philosopher, critic, and theologian – all were one in the unity of his intellectual enterprise. Science, through its foundation in facts and its informing structure of ideas and laws, related mind to nature, the ideal to the real, and had to be incorporated into his system. Science, in short, was fundamental in Coleridge's thought.

Coleridge, recognizing the creativity inherent in scientific discovery, saw in science a source of imaginative insight. And in interpreting science philosophically, as he had to do in striving for a true system, he drew upon the same philosophical canons that he used in moral philosophy. This took him to authors little regarded by English men of science; consider, for example, the unlearned castigation meted out to the German natural philosopher J. W. Ritter by Humphry Davy, then Coleridge's friend and England's leading scientist: "Ritter's errors as a theorist seem to be derived merely from his indulgence in the peculiar literary taste of his country, where the metaphysical dogmas of Kant which as far as I can learn are pseudo platonism are preferred before the doctrines of Locke and Hartley, excellence and knowledge being rather sought for in the infant than in the adult state of the mind."[9] And Davy in the same lecture condemned Plato for "hiding philosophy in a veil of metaphysical tinsel fitted only to pamper the senses." Such philosophical tastes were representative of those held by leading English scientists contemporary with Coleridge; his ideas about science were accordingly uncongenial to members of the scientific establishment. This establishment has until recently been virtually the exclusive preoccupation of historians of science. There was, however, in the early nineteenth century, a growing concern with exploring the interconnections and interfaces between the sciences; the German tradition of philosophical science was compatible with this unifying enterprise. The subsequent rejection of *Naturphilosophie* was emphatic; Justus von Liebig in the 1840s condemned it as "the black death of our century." The historiography of science followed the tenor of that rejection, so that we have largely ignored the extent to which German science was permeated by philosophy; we need to recognize the influence of *Naturphilosophie* even on its most influential scientific critics.[10] Coleridge, in working toward his system, drew widely on this alternative tradition. Now, when questions about science and romanticism are receiving attention, when the social history of science is maturing, and when many historians are relaxing an earlier positivism in their definition of science, Coleridge's views about science and his sources in science have a renewed significance.[11] At the same time, Coleridge's writings about science, previously difficult of access, are rapidly becoming available in successive volumes of the *Notebooks* and the *Collected Coleridge*.[12]

The sciences were prominent in Coleridge's earliest educational schemes, valuable in themselves and as an aid to the mind in perceiving relations and grasping ideas, essential parts of the poet's business. Coleridge knew what was needed as preparation for the writing of an epic poem – a grasp in principle and in detail of the knowledge of the ages, the history and frame of man and nature. And then the poet's mind would work upon this knowledge, transmuting it into a unity that mirrored nature through the synthetic and creative power of imagination. Poet, philosopher, and scientist were one in this enterprise. Coleridge saw Shakespeare as a nature humanized, poets as profound metaphysicians, and Humphry Davy's chemistry as poetry realized in nature.

Coleridge was a brilliant observer of the minutiae of nature. He perceived and recorded details, while seeking to comprehend their significance through their interrelations within the web of nature. He had a native ecological sense. At the same time, he saw clearly that relations were mental constructs, and no less real for that. He regarded nature as determined, and even defined it as the chain of cause and effect. This determinism was not mechanical, but dynamic, governed by powers. There were powers of mind and powers in nature, corresponding to one another. This network of correspondences made nature one, and made it intelligible.

In Coleridge's account, the human mind created unity through ideas, whereas nature's unity arose from laws. But there was a sense in which ideas were laws, so that philosophers could move from mind to nature, and scientists could move from nature to mind. The investigations of scientists were thus integral to Coleridge's lifelong inquiry into the rule of the active mind. It is significant that when he came in *The Friend* to illustrate right intellectual method, he did so with examples drawn from science and from the history of science. His illustrations of genius, the supreme sustained exertion of imagination, came as often from science as from literature. Imagination, for poet and scientist alike, transmuted and unified thought and thing, making mind one with nature. "The *rules* of the *IMAGINATION* are themselves the very powers of growth and production."[13] It follows, as Owen Barfield has remarked, "that anyone who has decided to take Coleridge seriously will be shirking the issue if he fails to consider the relation between what he thought, on the one hand – and 'Science' on the other."[14] What is important here is the way in which Coleridge thought about science, and the role of scientific information in the development of his thought; we concern ourselves with what Coleridge thought about science in order the better to grasp how he thought.

Science was valuable for Coleridge because it revealed and constituted relations in nature. It was the antidote to speculation in philosophy, and he used it accordingly. Philosophy was supported by science, to which it gave structure. "A system of Science *presupposes* – a system of Philosophy"[15] It therefore comes as no surprise to find that in the various drafts and schemes of Coleridge's great work of systematic intellectual synthesis, his unfinished *Opus Maximum*, science enters early and fundamentally. "The Logosophic System and Method . . . first demonstrating the inherent imperfection of all exclusively intellectual . . . or theoretical Systems, . . . proceeds to establish the ~~true~~ proper . . . character and function of Philosophy as the supplement of Science, and the realization of both in Religion or the Life of Faith."[16]

The *Opus Maximum* epitomizes Coleridge's whole intellectual enterprise. There were always plans for the work, which were always in the process of modification, but never came satisfactorily to fruition. There is method in the partial drafts, an articulation of ideas and arguments, a guiding structure; but for all his striving, one is hard put to discover system in them. The "collisions of a hugely developed sense of inner reality with a hugely developed sense of outer reality, with neither sense giving ground," produced an unresolved and, as McFarland has argued, an unresolvable tension between the philosophical traditions of "Platonico-Christo-Kantism on the one hand, and Spinozism, on the other."[17] Coleridge was led to the development of an argument, not the statement of a conclusion. He was frustrated in his attempts to create a system of philosophy; but he philosophized, and sought to teach others to do likewise. Scientific thought furnished an example of intellectual method, and thinking about science – what M. H. Abrams has called metascience[18] – was an essential part of Coleridge's critical philosophical and imaginative activity.

I have emphasized the tentative and progressive form of his study of mind and nature, with its incorporation of facts and laws of science. Coleridge, always his own worst advocate, seems to invite criticism for embarking on programs that could not be completed. But there is much in common between his enterprise and the enterprise of science, although the latter has often been conducted by those least patient with philosophizing. It is not merely that both go to nature and to mind in bringing intellectual order to sensory multiplicity and chaos. It is not even that some scientists, like some philosophers, impose their ideas on nature as a step toward finding them in nature. Science, like Coleridge's thought, is progressive, always perceiving and incorporating new facts, new laws and ideas. Science is also tentative, for new discoveries, new concerns, new ways of seeing the natural world, lead to the rejection of old theories, the modification of old laws, and the formulation of new ones. Again, science is always unfinished, partial, and selective. Coleridge, although no scientist, could sympathize with the work of scientists, and so study their writings the more eagerly.

Coleridge's exploration of the sciences and the formulation of his metascience were major components in the articulation and development of his thought. The intensity of his exploration is revealed in extensive evidence, in letters, notebooks, and marginalia, fragments indicative of a far-from-fragmentary intellectual grasp. Beddoes and Blumenbach, Darwin and Davy are prominent in the early years. Coleridge knew them personally and through their works. He studied physiology in Göttingen and chemistry at Davy's lectures at the Royal Institution in London, and he acquired besides a good general grounding in the science and medicine of his day. He contributed directly and indirectly to the lectures of Joseph Henry Green and tackled physiological problems with Dr. James Gillman. German philosophical science fascinated him, and he discussed science with visitors like Friedrich Tiedemann and G. R. Treviranus. He always retained his interest in science. In 1833, one year before his death, he visited Cambridge to attend a meeting of the British Association for the Advancement of Science, characteristically contributing to a debate on the proper title for men of science.

Coleridge's response to intellectual issues was never wholly abstract. Just as nature was his touchstone in metaphysics, so his scientific friends and teachers disciplined and directed his approach to science. He shared their enthusiasms wholeheartedly. Davy's electrochemical investigations and his researches in animal chemistry were accordingly of seminal importance for Coleridge.

He was well informed about most of the sciences of his day – more so than most philosophers of science have been before or since, and perhaps uniquely so for a poet. The range and quality of his metascience bespeaks an involvement intense because interdependent with his other commitments in poetry, theology, and criticism. He acquired real familiarity with the theory and factual content of essentially qualitative sciences like physiology, chemistry, and geology. Mathematical physics and astronomy were, however, largely closed to him by his ignorance of mathematics. This imbalance in his knowledge was troublesome to him, for he aimed at a comprehensive account of the sciences as part of his unified encyclopedic study of man and nature and of their relation to God.

He believed that the difficulty could be overcome. He regarded quantitative sciences as essentially analytical, deriving from Newton's physics and Locke's philosophy. His own scheme was to be synthetic, exhibiting relations through polar powers, so that the quantitative aspects of science were of only secondary importance to him. The grand synthesis always escaped him. It was, like final truth in science, unattainable, even had some areas of science not been closed to him. He never published his *Opus Maximum*, although he wrote or dictated several volumes of it. He never wrote a comprehensive philosophy of nature. But for a decade after 1816, he worked steadily toward one. The *Theory of Life*, his most extended metascientific statement, may be seen as a step toward the desired system.

System implies and demands method. Coleridge's method was to seek fundamental relations and correspondence in the light of ideas. His fundamental ideas were the Trinity, and polarity, which he saw as its corollary. Polarity, as Barfield has written, "is dynamic, not abstract. It is not 'a mere balance or compromise,' but 'a living and generative interpenetration' . . . [T]he apprehension of polarity is itself *the basic act of imagination*."[19]

Polarity was the crucial concept in Coleridge's dynamic logic, elaborated in response to his need to reconcile the existence of man's moral nature and the transcendence of God with the reality of external nature. It was part of his trinitarian resolution[20] of the most fundamental problems, reached at the end of his passage from Unitarianism and necessitarianism, always developing and never fully elaborated. This progress took him not only to Kant, whom he was to read until his death, but to F. W. J. von Schelling and Henrik Steffens, who used polarity in an attempt to reconcile subject and object, the self and nature. Coleridge first accepted their arguments, then perceived the pantheistic implications of a philosophy that made nature absolute. He then moved away from Schelling and Steffens, transcending rather than rejecting them, often preserving their language while redefining crucial terms.

His use of German sources in the years immediately following the composition of the *Biographia Literaria* is as significant in metascience as in other interdependent regions of his thought; here as elsewhere, Coleridge used his sources in ways that demonstrate understanding of the issues they confronted, moving frequently and perceptively beyond them. He used the facts of science, drawn from impeccable researches and reliable compendia, together with the ideas of philosophers, in working toward his own system.[21] His thought about science, developing with his philosophy and theology, can be studied through his changing views of *Naturphilosophie* in the decade after 1815.

Coleridge believed that mind was active in nature, which was itself organic, alive and developing, and intelligible. Here were the foundations for a systematic study. In the years after his move to Highgate, he was in almost daily intercourse with medical men, and became increasingly interested in the activities of the Royal Institution, the Royal Society, the Society for Animal Chemistry, and the Royal College of Surgeons. He pursued a program of scientific reading for several years from 1819. His notebooks reveal a striking constancy of purpose, with methodical attention devoted to some areas of study, especially chemistry. His wide readings in English and German scientific texts broadly followed the hierarchy of the natural sciences that was to be incorporated in the *Opus Maximum*.

Coleridge followed the Bible in moving from chaos through the construction of matter to the construction of the cosmos. The Bible, after all, was "in its own way a mythic and holy book written in symbols that explained creation, nature, God, and man in a poetic language."[22] Coleridge's definition of the primary imagination as "a repetition in the finite mind of the eternal act of creation in the infinite I AM"[23] related creation, God, the self, and nature in a unity that could be explored through the biblical account of creation, illuminating the creation and the order of nature. There could be no conflict between revelation and fact, no temptation toward lying for God. "If Christianity is to be the religion of the world . . . so true must it be that the book of nature and the book of revelation, with the whole history of man as the intermediate link, must be integral and coherent parts of one great work: and the conclusion is, that a scheme of the Christian faith which does not arise out of, and shoot its beams downward into, the scheme of nature, but stands aloof as an insulated afterthought, must be false or distorted in all its particulars."[24]

Science, philosophy, and religion provided the foundation, with the facts of science at once a perpetual touchstone and the fabric of the edifice. Coleridge, incorporating the latest and surest discoveries of science, moved from cosmology through astronomy, physics, geology, and chemistry to the life sciences. All his scientific reading, his medical gleanings, and his interest in natural history combined with philosophy to give him a theory of life. Coleridge succeeded in his notebooks in constructing an approach to the sciences that welded their parts into a unity, and offered a radical alternative to the scientific orthodoxy of his time. The scheme, unfinished as it is, adds a new dimension to our understanding of Coleridge's thought and of early nineteenth-century science.

1

EARLY YEARS

FROM HARTLEY TO DAVY

Hartley and Priestley

Politics and religion provided Coleridge's introduction to science.[1] In December 1795 he had published a sonnet to Joseph Priestley, whose sympathy for the French Revolution had wrought the mob to drive him out of Birmingham. Priestley was a Socinian,[2] and thus congenial to Coleridge, then a Unitarian and a Democrat.[3] And Priestley had recently crossed the Atlantic in search of freedom from persecution. In his *Experiments and Observations on Different Kinds of Air*, he had stated his creed as a natural philosopher:

> This rapid process of knowledge, which, like the progress of a wave of the sea, of sound, or of light from the sun, extends itself not this way or that way only, but *in all directions*, will, I doubt not, be the means, under God, of extirpating *all* error and prejudice, and of putting an end to all undue and usurped authority in the business of religion, as well as of *science*; and all the efforts of the interested friends of corrupt establishments of all kinds, will be ineffectual for their support in this enlightened age; though, by retarding their downfall, they may make the final ruin of them more complete and glorious. It was ill policy in Leo X. to patronize polite literature. He was cherishing an enemy in disguise. And the English hierarchy (if there be anything unsound in its constitution) has equal reason to tremble even at an air pump, or an electrical machine.[4]

Coleridge seized on this message in his "Religious Musings: A Desultory Poem, Written on the Christmas Eve of 1794":

> From Avarice thus, from Luxury and War
> Sprang heavenly Science; and from Science Freedom.

Priestley, driven abroad, was "patriot, and saint, and sage," prophet of reformation and regeneration.[5] His *Disquisitions on Matter and Spirit* entered significantly into the poem.[6] Coleridge's tribute to Priestley was preceded here by tributes to Isaac Newton and to David Hartley, a physician and former fellow of Coleridge's Cambridge college.[7] Hartley combined physiology and psychology with a corresponding validation of free will and religion, basing his theory on the works of Newton and Locke. This attempted reconciliation of faith and reason, however narrowly conceived, was attractive to Coleridge in his desire for an intellectual foundation for duty and faith.

Hartley also gave a philosophical foundation to one of Coleridge's favorite arguments for poetry, that it freed man from the tyranny of the senses. Hartley had begun by asserting that man consisted of two parts, body and mind. Then, working from hints in Newton's *Principia* and in the queries at the end of his *Opticks*, and from Locke's doctrine of association, Hartley proposed his doctrine of vibrations. Vibrations in the ether were transmitted by human organs of sensation to the brain, and thus immediately to the mind, where, if they recurred, they left "*certain Vestiges, Types, or Images of themselves, which may be called,* Simple Ideas of Sensation." Simple ideas could run into complex ones by means of association, and complex ideas could be as vivid as simple ones.[8] Coleridge, gratified, and not yet having defined what he meant by an idea, wrote of this account: "Ideas may become ~~so~~ ⟨as⟩ vivid & distinct; & the feelings accompanying them as vivid, as original impressions – and this may finally make a man independent of his senses."[9]

Hartley managed uneasily to salvage free will in his theory of association by making a distinction between philosophical free will, which he considered impossible, and free will in its popular and practical sense, "a voluntary power over our Affections and Actions." As philosophical free will was the power of doing different things when previous circumstances were unchanged, Hartley was a necessitarian. Coleridge fully concurred: "I am a compleat Necessitarian – and understand the subject as well almost as Hartley himself – but I go farther than Hartley, and believe the corporeality of *thought* – namely, that it is motion."[10]

Thomas Beddoes and Erasmus Darwin

Coleridge's political opinions and his friendship for Robert Southey had brought him from Cambridge and London to Bristol.[11] Thomas Beddoes had forsaken Oxford for Bristol eighteen months previously, in part because his open sympathy for the French Revolution had aroused friction in Oxford, in part because attendance at his chemistry lectures and thus his stipend there had fallen, and in part because of the reaction to his *Alexander's Expedition*, a printed but unpublished epic poem denouncing British imperial ambitions in India.[12] Bristol was attractive as a lively radical center and, thus far, had similar attractions for Beddoes and Coleridge. In 1795 they both attended a public meeting to protest against the Pitt–Grenville "Gagging Bills"; this may have been when they first met. They saw much of one another in Bristol beginning in 1795, and Coleridge's interest in science significantly dates from the same year.[13] Beddoes, whose ideas were seminal for Coleridge's scientific development, was responsible for his meeting Davy; Davy's influence on Coleridge persisted even after their estrangement.

Beddoes's motives in moving to Bristol had not been merely political. He was, after all, a medical doctor who taught chemistry at Oxford and had studied under Joseph Black in Edinburgh. In particular, he had a long-standing interest in the chemistry of airs, was familiar with the writings of Carl Wilhelm Scheele and Priestley, and in 1790 had edited and published extracts from John Mayow's writings, with reflections on the chemistry of respiration.[14] At the same time Beddoes's work as a doctor regularly confronted him with consumptive patients. Because airs interacted with the blood in the lungs, and because consumption was a disease of the lungs, he thought of using pneumatic chemistry to treat consumption. In a public letter to Erasmus Darwin, Beddoes explained that he had fixed in Bristol Hotwells to pursue this problem, "because this resort of invalids seemed more likely than any other situation to furnish patients in all the various gradations of Consumption." He set about raising funds for a Pneumatic Medical Institution.[15]

Beddoes, corresponding with Darwin about his practice and proposals and admiring Priestley's discoveries, was also treating Thomas Wedgwood's and Gregory Watt's illnesses,[16] and working with James Watt to develop a pneumatic apparatus for administering gasses to patients. Completing his ties with the members of the Lunar Society, Beddoes became engaged to Anna Maria, daughter of Richard Lovell Edgeworth. Edgeworth observed of his future son-in-law that he was "a little fat Democrat of considerable abilities, of great name in the Scïentific world as a naturalist and Chemist – good humor'd good natured – a man of honor & Virtue, enthusiastic & sanguine. . . . His manners are not polite – but he is sincere & candid . . . if he will put off his political projects till he has accomplish'd his medical establishment he will succeed and make a fortune – But if he bloweth the trumpet of Sedition the Aristocracy will rather go to hell with Satan than with any democratic Devil."[17] Edgeworth's assessment was shrewd. Beddoes blew the trumpet of sedition with vigor, arousing the alarmed suspicion of government officials, antagonizing the scientific establishment ruled by Joseph Banks and the Royal Society,[18] and captivating Coleridge, heart and head.

Their enthusiasm was mutual. When Coleridge produced his miscellany, *The Watchman*, to appear every eight days starting on February 5, 1796, Beddoes subscribed, contributed, and was reviewed by Coleridge: "To announce a work from the pen of Dr. Beddoes is to inform the benevolent in every city and parish, that they are appointed agents to some new and practicable scheme for increasing the comforts or alleviating the miseries of their fellow-creatures."[19]

Political reform and scientific advance could both lead to social amelioration. Coleridge, subscribing himself "as a faithful / WATCHMAN / to proclaim the State of the Political Atmosphere, and preserve Freedom and her Friends from the attacks of Robbers and Assassins!!" saw science as a friend of freedom. In the first number, he seized on the passage in Burke's "Letter to a Noble Lord" that attacked the disposition of metaphysicians, geometricians, and chemists. Any system that could reckon such men among its enemies was for Coleridge essentially vile. "But the sciences suffer for their professors; and Geometry, Metaphysics, and Chemistry, are Condorcet, Abbe Sieyes, and Priestley, generalized."[20]

Beddoes was clearly a natural subscriber to *The Watchman*. He was in regular correspondence with Erasmus Darwin in the early and mid-1790s, about medicine, physiology, geology, and philosophy.[21] Darwin had sent Beddoes the manuscript of *Zoonomia* prior to its publication, and Coleridge's notebooks make it clear that in 1795 he became acquainted with the ideas in *Zoonomia* and read Darwin's *Botanic Garden* with some care. The relevant notebook entries were written shortly after Coleridge and Beddoes first met. It is likely that Beddoes, here as in so many cases, encouraged Coleridge's scientific enthusiasm, although Coleridge's interest in Darwin had probably been aroused by discussions at Cambridge in 1793.[22]

In late January 1796, Coleridge reported his meeting with Darwin to his Bristol friend Josiah Wade:

> Derby is full of curiosities, the cotton, the silk mills, Wright, the painter, and Dr. Darwin, the everything, except the Christian! Dr. Darwin possesses, perhaps, a greater range of knowledge than any other man in Europe, and is the most inventive of philosophical men. He thinks in a *new* train on all subjects except religion . . . Dr. Darwin would have been ashamed to have rejected Hutton's theory of the earth without having minutely examined it; yet what is it to us *how* the earth was made, a thing impossible to be known, and useless if known? This system the

> Doctor did not reject without having severely studied it; but *all at once he makes up his mind* on such important subjects, as whether we be the outcasts of a blind idiot called Nature, or the children of an all-wise and infinitely good God.[23]

Coleridge was later to reject Darwin's poetry and his science. But in the mid-1790s, Darwin seemed to him to be, on the whole, "the first *literary* character in Europe, and the most original-minded man." Darwin was certainly a man of wide and various knowledge. The notes to his extended scientific poems were a treasure trove, introducing Coleridge to the latest works in chemistry, meteorology, natural philosophy, and geology.[24]

Beddoes, meanwhile, seems to have given Coleridge access to his extensive library of scientific works. The library was up-to-date, containing not only the standard British works, but also a range of French and, more strikingly, of German scientific and philosophical texts and periodicals. When Humphry Davy was ensconced in the Pneumatic Institution, Gregory Watt wrote to him: "You have as early access as myself to foreign journals."[25] Some indication of the size of Beddoes's collection may be derived from the account given by Dr. Joseph Frank of Vienna of his visit to Beddoes in 1803. Frank sent in his name, and after a while Beddoes came in, burdened with an armful of volumes, each by a different Dr. Frank. Joseph Frank recalled that the conversation turned to "foreign medical literature; when I soon found that Dr. Beddoes reads German as well as he does English; and is intimately acquainted with all our best authors."[26] A similar conclusion could be drawn from Beddoes's criticisms of the Bodleian Library in his Oxford days,[27] or from the reviews he published. The *Monthly Review*, from Coleridge's arrival in Bristol in 1795 to his departure for the Lakes in 1800, carried 136 reviews by Beddoes, mostly of medical and chemical works, but extending also to philosophy and anthropology. Of these reviews, 4 were of Italian works, 18 of French works, and 32 of German works. These last included works by Kant, Johann Friedrich Blumenbach, Christoph Heinrich Pfaff, Christoph Wilhelm Hufeland, and Johann Christian Reil.[28] Coleridge did not immediately seize on these authors – his lack of German would have prevented him from doing so. There is, however, an extensive overlap between Beddoes's German review list and Coleridge's scientific reading as it appears in the notebooks of his productive middle age.

Meanwhile, Coleridge's scientific reading was largely restricted to English and Latin texts. In 1795 and 1796 his letters and notebooks show that among the books he read were Darwin's *Botanic Garden*;

Priestley's *Opticks* and, most probably, his *Disquisitions on Matter and Spirit*; Count Rumford's *Essays*; Newton's works (including the *Opticks*, the *Letter to Mr. Boyle*, and the *Principia*, much of which must have been unintelligible to him); and a work or works by Beddoes and Watt on pneumatic medicine. He also read articles in journals, including John Haygarth's account of a "Glory" in *Memoirs of the Literary and Philosophical Society of Manchester* and, according to J. L. Lowes, John Hunter's "Observations on the structure and oeconomy of whales" in the *Philosophical Transactions of the Royal Society of London* for 1787. Beddoes's library appears to have supplied the monographs; Coleridge borrowed the periodicals from the Bristol Library Society.[29]

Thought and feeling

Science at first formed a relatively insignificant part of Coleridge's omnivorous reading. As the "library-cormorant" wrote late in 1796, "Metaphysics, & Poetry, & 'Facts of mind' . . . are my darling Studies. – In short, I seldom read except to amuse myself – & I am almost always reading. – Of useful knowledge, I am a so-so chemist, & I love chemistry – all else is *blank*."[30] Coleridge so far had had an erratic exposure to eighteenth-century British science and philosophy. Newton, Hartley, and Priestley together could be made to furnish a philosophical foundation for Unitarianism and for libertarian politics. Soon, however, as Coleridge came to stress the active mind and creative imagination in his philosophy, he abandoned Hartley's necessitarianism. His growing knowledge of German literature, his own sentiments, and the demands made by their expression were reinforced by his religious feelings (at this point scarcely a theology), and now they all began to bear down on him, inculcating frustration with a whole tradition of science and philosophy. John Locke's epistemology, deriving knowledge primarily from the sense of sight, began to seem superficial and incoherent. A mechanical lifeless world of passive atoms and imponderable fluids was empty of meaning, fragmented, and godless. In "The Destiny of Nations," written in 1796, Coleridge attacked the emptiness of scientific materialism and proposed a living alternative for science and poetry:

> For what is Freedom, but the unfettered use
> Of all the powers which God for use had given?
> But chiefly this, him First, him Last to view
> Through meaner powers and secondary things
> Effulgent, as through clouds that veil his blaze.
> For all that meets the bodily sense I deem

> Symbolical, one mighty alphabet
> For infant minds.

Natural phenomena were symbols of something higher, something conferring unity and life and significance on what would otherwise be meaningless and disconnected. Science and poetry, complementing one another, could help mankind "to know ourselves / Parts and proportions of one wondrous whole!" Science in its eighteenth-century empiricist guise had neglected this role:

> But some there are who deem themselves most free
> When they within this gross and visible sphere
> Chain down the wingéd thought, scoffing ascent,
> Proud in their meanness: and themselves they cheat
> With noisy emptiness of learnéd phrase,
> Their subtle fluids, impacts, essences,
> Self-working tools, uncaused effects, and all
> Those blind Omniscients, those Almighty Slaves,
> Untenanting creation of its God.[31]

"Untenanting creation of its God" – here was a crisis of religion and of feeling, Coleridge believed that this had been brought about by unthinking adherence to Locke's system of philosophy, which he held increasingly in contempt. The interfusion of thought with feeling was central to Coleridge's being. "I feel strongly," he wrote, "and I think strongly; but I seldom feel without thinking, or think without feeling . . . My philosophical opinions are blended with, or deduced from, my feelings." Coleridge was not advocating the rule of feelings in science, where facts were always a touchstone. But he wanted the philosophy with which he approached those facts to be in accord with his feelings. He felt that there was an all-encompassing unity; empirical philosophy, founded in the senses, gave him instead "an immense heap of *little* things."[32] He was reacting against the narrow rationality of the Enlightenment, and instead of Rousseau admired Bishop Berkeley and David Hartley, and went back to such seventeenth-century divines as Jeremy Taylor and Richard Baxter.[33] Here was a revolt away from mere empiricism, and the first step toward idealism in philosophy and trinitarianism in religion. And just as Priestley's natural philosophy supported Unitarianism, so another style of natural philosopher would later be appealed to in support of Coleridge's developing ideas about nature, mind, and God. Something was needed to replace mere empiricism.

Coleridge already admired Schiller's dramatic verse, and knew of accounts Beddoes had written of Kant's philosophy and of German science – knew, indeed, that Beddoes believed that the palm of science and literature belonged to the country of Haller, Heyne, Meiners, and Michaelis, of Doederlein, Reimarus, Mendelssohn, and Lessing, of Bloch, Jacquin, Pallas, and Schreber. And for a "library cormorant," the superiority that Beddoes had claimed for the university library at Göttingen would have had an unanswerable attraction.[34] Coleridge became persuaded that he needed to go to Germany.

Schemes of education: to Germany

In May 1796, Coleridge was alternately despondent and ebullient as he looked into his recent past and imagined future. To Tom Poole, tanner, democrat, and by now a regular confidant, he set out his plans,

> reduced to two –. The first impracticable – the second [to become a Unitarian minister] not likely to succeed – Plan 1st. – I am studying German, & in about six weeks shall be able to read that language with tolerable fluency. Now I have some thoughts of making a proposal to Robinson, the great London Bookseller, of translating all the works [of] Schiller, . . . on the conditions that he should pay my Journey & wife's to & from Jena, a cheap German University where Schiller resides – & allow me two guineas each Quarto Sheet – which would maintain me –. If I could realize this scheme, I should there study Chemistry & Anatomy, [and] bring over with me all the works of Semler & Michaelis, the German Theologians, & of Kant, the great german Metaphysician. On my return I would commence a School for 8 young men at 100 guineas each – proposing to *perfect* them in the following studies in order as follows –
>
> 1. Man as Animal: including the complete knowledge of Anatomy, Chemistry, Mechanics & Optics. –
> 2. Man as *Intellectual* Being . . .
> 3. Man as a Religious Being . . . History . . . [35]

This scheme was neither immediately nor fully realized. But Coleridge's view of the structure of right education was already formed in essentials. First came the sciences, the study of nature including man as object; then came philosophy, which embraced the study of man as subject; finally came the study of religion and of the social sciences, incorporating the history of science and of philosophy. Coleridge, still a Unitarian, regarded every experiment that Priestley made in chemistry as "giving *wings* to his more sublime theological works." Science and its history would reinforce religion. The whole educational scheme outlined here would strengthen intellect in the service of religion, and provide the "*foundation* of excellence in all professions."[36]

Charles Lamb, whose taste and judgment at this time Coleridge thought "more correct and philosophical than my own, which yet I place pretty high,"[37] urged Coleridge to write an epic poem. Coleridge considered such a project, and found in his educational ideas good reason for deferring it. He wrote to Joseph Cottle, his publisher in Bristol:

> Observe the march of Milton – his severe application, his laborious polish, his deep metaphysical researches, his prayers to God before he began his great poem, all that could lift and swell his intellect, became his daily food. I should not think of devoting less than 20 years to an Epic Poem. Ten to collect materials and warm my mind with universal science. I would be a tolerable Mathematician, I would thoroughly know Mechanics, Hydrostatics, Optics, and Astronomy, Botany, Metallurgy, Fossilism, Chemistry, Geology, Anatomy, Medicine – then the *mind of man* – then the *minds of men* – in all Travels, Voyages and Histories. So I would spend ten years – the next five to the composition of the poem – and the five last to the correction of it.[38]

The program was impossibly ambitious, yet Coleridge went a good way toward its achievement. And still studies in Germany seemed the necessary prelude to the program. But how was he to go there, with a family to support and scarcely a secure income?

Help for Coleridge came from the Wedgwoods. Thomas Wedgwood had been under treatment – several courses of treatment – by Beddoes, with little success. Wedgwood, who was gifted in metaphysical discussion and fascinated by the natural sciences, found himself in full sympathy with Coleridge. He and Josiah, recognizing Coleridge's abilities and his needs, gave him a pension. In June 1798, Coleridge visited his patrons at Stoke d'Abernon in Surrey and discussed his plans with both of them.[39]

He and the Wordsworths sailed from Yarmouth on September 16, arriving in Hamburg three days later.

Coleridge went walking, learned some German, separated from the Wordsworths and went to Ratzeburg to be better placed to learn more German, dined all around, marveled at German pipe smoking, and talked metaphysics. He met "every where" with Kantians – "SNUFFS that have a live spark in them – & fume under your nose in every company." Kant, whom he had formerly dismissed in frustration as "the most unintelligible Emanuel Kant," had since become important for his plan of study. He failed to master Kant while in Germany, although he seems to have intended to try.[40]

On February 6, 1799, he left Ratzeburg for Göttingen, where he matriculated at the university and visited the library. Beddoes had not exaggerated. It was "without doubt . . . the very first in the World both in itself, & in the management of it."[41] Coleridge's studies in the ensuing weeks included attendance at Blumenbach's lectures. Beddoes knew Blumenbach's work; he had reviewed his book on anthropology and drawn attention to his textbook on physiology.[42] Blumenbach was Germany's foremost teacher of physiology and natural history. Coleridge attended his lectures morning and evening, enjoyed his studies, and remembered what he learned.[43]

Coleridge's enthusiasms in science were at this date still for his friends' concerns. He made notes about Brunonian physiology, one of Beddoes's interests, and wrote to Tom Poole about beet sugar.[44] But it seems clear from the paucity of scientific entries in the notebooks that Coleridge's plans for methodical self-education in science were premature. Indeed, his only systematic exposure to natural science in Germany came from Blumenbach's lectures.

Clement Carlyon, one of Coleridge's fellow students at Göttingen, tells how in May 1799, after a spell of close application to their academic studies, a group of Englishmen including Coleridge and himself, accompanied by one of Blumenbach's sons, set off for a walking tour in the Harz mountains. George Bellas Greenough was another of the party. He had arrived in Göttingen in 1798 to study law, had come under Blumenbach's influence, and had subscribed to his lecture course in October. Like Coleridge at this date, he also had an amateur's interest in chemistry. He sat next to Coleridge in Blumenbach's lectures on physiology and in 1799 was developing the interest in geology and mineralogy that was to lead him to become first president of the Geological Society in 1811. He introduced Coleridge to geology, showed him his collection of minerals in Göttingen, and ensured that a copy of René-Just Haüy's *Traité de Minéralogie* was in Coleridge's baggage on the voyage to Malta in 1804.[45]

The *Hartzreise* was more than an introduction to geology (and to the "glory" of the Brocken specter)[46] – it was a walking tour in the mountains and was relished by Coleridge, who practically invented mountaineering as literally a re-creation. On this tour he looked down into the Vale of Rauschenbach, walked on – "And now on our left hand came before us a most tremendous Precipice of [?y]ellow & black Rock, called the Rehburg . . . – Now again is nothing but Pines & Firs, above, below, around us! – How awful is [?the] deep Unison of their undividable Murmur – What a *one* thing it is [?– it is a sound] that [?im]presses the dim notion of the Omnipresent!"[47] His perception of nature embraced equally the minutiae of individual things and the unity of all things. This polarity, fundamental to his perception and to his feelings, interpenetrated his thought, and so is part of the texture of his finest poems and also of his use of scientific information. Each fact of experience had to be appreciated in itself, and simultaneously drawn into the intellectual unity of nature whose comprehension and demonstration was the scientist's and the poet's goal. Coleridge, inspired by the chemical researches of his contemporaries, would later eloquently advance a fine complementarity in poetic and scientific creativity.[48] For the present, in May 1799, he was walking in the mountains after a month of intensive study, finding a dynamic, rhythmic unity of design in the prospect of the hills.

Coleridge's studies in Göttingen were over. He reported his achievements to Josiah Wedgwood in a letter of May 21, 1799. He had learned to read high and low German and to speak the former "so fluently, that it must be a *torture* for a German to be in my company – that is, I have words enough & phrases enough, & I can arrange them tolerably; but my pronunciation is hideous" – a diagnosis confirmed for us by notebook entries.[49] He had attended lectures on physiology, anatomy, and natural history, and had "endeavoured to understand these subjects." Blumenbach's lectures were widely esteemed and seem to have been as admirably organized and lucid as his textbooks. He had, moreover, the art, Carlyon tells us, of bringing the peculiarities of animals before his audience, "by tone or by gesture, . . . in a more lively manner than mere verbal description would have done," without derogating from his professional dignity. Besides engaging in his linguistic, scientific, and literary studies, Coleridge had bought "30 pounds worth of books (chiefly metaphysics / & with a view to the one work, to which I hope to dedicate in silence the prime of my life)." These books were to lend a vocabulary and structure to the thought that directed his scientific interests. They also attracted him to a range of sources and ideas unfamiliar to most of his countrymen, and generally uncongenial when familiar to them. Coleridge had still only a slender foundation in science, but he had the materials that would furnish him with method and bring coherence to subsequent scientific reading. The voyage to Germany marked a beginning, and gave a bearing to his studies.[50]

Coleridge left Göttingen on June 24, 1799, after a lively farewell party given by Blumenbach. At Brunswick on June 30, Coleridge and the English group met and talked with Professor Zimmermann, who criticized Kant and claimed that most German literati were Spinozists. Next came a visit to Christian Wiedemann, professor of anatomy, chemistry, and mineralogy, who showed them his museum with its excellent cabinet of minerals. At Brunswick, too, Coleridge inspected the Duke of Brunswick's cabinet of minerals.[51] And from Brunswick,

Coleridge, anxious for his chest of books and anxious to be home, made for Hamburg; he was back in Stowey by the end of July. Shortly afterward, he met Davy.

Humphry Davy: chemistry, 1799–1804

Humphry Davy had come to Bristol in October 1798 as chemical superintendent of Beddoes's Pneumatic Institution. He was full of effervescent enthusiasm, tireless conversation, and ambition well founded on a belief in his own genius. In his first surviving notebook, dating from 1795, Davy had set out a plan of study, like Coleridge's as revealing as it was impossible of realization.[52]

Robert Southey had encouraged Davy to write poetry. Davy responded vigorously, sending Southey plans and drafts of poems for criticism, and submitting poems for publication in Southey's *Annual Anthology*.[53] The first volume of this would-be annual contained Davy's "Sons of Genius," boldly expressing his personal and scientific aspirations. It is perhaps as well that Davy did not bring to the test claims that, had he not been the greatest scientist of his age, he would have been the greatest poet.[54] But although he later dismissed all thoughts of becoming a poet as the dreams of his youth, he took the liveliest interest in the writings of Southey, Coleridge, and Wordsworth, and, indeed, in his own poetic experiences.[55] Fresh from intercourse with Coleridge, he sought to discover and to share his friend's approach to nature and to analyze his own responses self-consciously. Davy's sense of personal sympathy with a unified and living nature was perhaps a little forced by a desire to emulate his poetic friends. But his view of nature was still very much his own: "Deeply and intimately connected are all our ideas of motion and life, and this, probably, from very early association. How different is the idea of life in a physiologist and a poet!"[56]

Clearly, Davy and Coleridge had complementary interests and wide and wild ambitions, and they were infectious in their enthusiasms. They impressed one another from the start, applauded, cheered, cajoled, and inspired each other. They talked together, walked together, worked together. They breathed nitrous oxide together in Beddoes's Pneumatic Institution. Coleridge felt a warm glow, and barely kept himself from laughter.[57] He attended Davy's attempts to treat patients with nitrous oxide. On one occasion, Davy was preparing to treat a patient in Coleridge's presence and, as a preliminary, put a thermometer under the patient's tongue. The patient, expecting a cure, immediately announced that he felt better. Davy then cast "an intelligent glance" at Coleridge, told the patient to return on the following day, and repeated the treatment daily for a fortnight. The patient was then dismissed as cured by the regular use of the thermometer alone. Davy and Coleridge were both interested in psychosomatic ailments, and Coleridge's correct interpretation of Davy's "intelligent glance" illustrates the closeness of the bond between them.[58] Chemistry, through Davy, became a subject of the greatest interest for Coleridge, and was to be central in his schemes of the sciences. Genius was of abiding interest to him, and Davy became his constant exemplar of scientific genius. And their joint interest in science and poetry stirred Coleridge to deeper inquiry into the relations between two realms of creative activity. Until disillusionment set in, Davy seemed to Coleridge the greatest man of the age after Wordsworth.[59]

In the autumn of 1799, Coleridge, in response to an invitation from Daniel Stuart, became a regular contributor to the *Morning Post.* Journalism, especially political journalism, occupied him through the winter and the spring of 1800. So did conversation and dining. Coleridge saw a good deal of William Godwin that winter. "I like him," he wrote to Thomas Wedgwood, "for thinking so well of Davy. He talks of him every where as the most extraordinary human Being, he had ever met with. I cannot say that: for I know one whom I feel to be the superior – but I never met so extraordinary a young man."[60] Coleridge's attachment to Davy was accompanied by a corresponding attachment to chemistry. His enthusiasm was compounded of admiration of the chemist and ignorance of his science. Through Beddoes, Coleridge had merely become aware of chemistry, in which he was very much a dilettante. His acquaintance with chemical theory came later. He wrote to Davy on New Year's Day, 1800, discussed the publication of Davy's researches; dreamed of a colony comprising Davy, James Tobin, Wordsworth, Southey, and himself; threw out questions about the philosophy of sensation and the nature of death; and then moved on to Godwin, chemistry, and poetry:

> Godwin talks evermore of you with lively affection – "What a pity that such a Man should degrade his vast Talents to Chemistry" – cried he to me. – Why, quoth I, how, Godwin! can you thus talk of a science, of which neither you nor I understand an iota? &c &c – & I defended Chemistry as knowingly at least as Godwin attacked it – affirmed that it united the opposite advantages of immaterializing [?the] mind without destroying the definiteness of [?the] Ideas – nay even while it gave clearness to them – And eke that being necessarily [?per]formed with the passion of Hope, it was p[?oetica]l – & we both agreed (for G. as

we[?ll as I] thinks himself a Poet) that *the Poet* is the Greatest possible character – &c &c. Modest Creatures![61]

Coleridge's philosophical and physiological inquiries had been prompted and directed by the recent publication of Davy's "Essay on heat, light, and the combinations of light" in Beddoes's collection of *Contributions to Physical and Medical Knowledge, Principally from the West of England.*[62] In this volume, Beddoes and Davy had both criticized Lavoisier's doctrine that chemists should consider the phenomena of heat as deriving from caloric, the "matter of heat," which he proposed as a simple substance. Views shared by Beddoes and Davy coincided with the implications of Rumford's cannon-boring experiments in undermining Lavoisier's doctrine. Lavoisier had acknowledged Locke and Condillac, and English chemists were beginning to acknowledge Lavoisier.[63] Here was a process that Coleridge, increasingly dissatisfied with eighteenth-century empiricism, had already identified and condemned in philosophy: "Hume wrote – and the French imitated him – and we the French – and the French us – and so philosophisms fly to and fro – in serieses of imitated Imitations – Shadows of shadows of shadows of a farthing Candle placed between two Looking-glasses."[64] Now in chemistry the whole absurd process was encountering opposition, Lavoisier's system was coming under attack, and what Beddoes and Davy saw as its artificial complexities were being rejected in favor of ideas about the chemical elements similar to Priestley's.[65]

Davy's "Essay" attacking Lavoisier was Newtonian, attributing powers of attraction and repulsion to matter, and denying the existence of substantial caloric. In Davy's scheme, light played a major role, entering into chemical combination and thus into the composition of living bodies. He proposed that light combined with oxygen to form phosoxygen. "On the existence of this principle in organic compounds, perception, thought, and happiness, appear to depend."[66] Davy discussed the probable action of light and oxygen in respiration and sensation, and drew wide-ranging conclusions. He suggested that life might be considered "a perpetual series of peculiar corpuscular changes," and that laws of mind were no different from the laws of corpuscular motion. Chemical and mechanical laws would then appear

as subservient to one grand end, PERCEPTION. Reasoning thus, it will not appear impossible that one law alone may govern and act upon matter: an energy of mutation, impressed by the will of the Deity, a law which might be called the law of animation, tending to produce the greatest possible sum of perception, the greatest possible sum of happiness.

> The farther we investigate the phaenomena of nature, the more we discover simplicity and unity of design. . . .
>
> We cannot entertain a doubt but that every change in our sensations and ideas must be accompanied with some correspondent change in the organic matter of the body. These changes experimental investigation may enable us to determine. By discovering them we should be informed of the laws of our existence, and probably enabled in a great measure to destroy our pains and to increase our pleasures.
>
> Thus would chemistry, in its connection with the laws of life, become the most sublime and important of all sciences.[67]

Newton, Hartley, and Priestley, gods in Coleridge's scientific pantheon, were here united in Davy's program – and perhaps their genius would be matched in his person. It was little wonder that Coleridge was enthralled. Davy was attempting an integration, however imbalanced, of the three components of the plan of education Coleridge had proposed in 1796 – man as animal, man as intellectual being, and, less prominently in Davy's scheme, man as a religious being. All would be brought together through the philosophy of perception.

Davy was, however, only the second greatest man of the age. In June, Coleridge wrote to Davy from Stowey announcing that he had finally decided that he would live in the north – in Keswick, near Wordsworth. He asked whether he should translate Blumenbach's *Naturgeschichte* (of which he possessed the latest edition), and again addressed himself to metaphysics. He asked Davy for his "metaphysical system of Impressions, Ideas, Pleasures, & Pains," in order to compare Davy's views with those of Spinoza and of Leibniz, which he planned to study as soon as he had settled in Keswick.[68]

Once in Keswick, Coleridge urged his invitation on Davy – "But you *will* come" – anxiously reminded him of his promise to send a synopsis of his metaphysical opinions, and assured him that he intended to "attack chemistry, like a Shark."[69] This union of science and metaphysics was fundamental to Coleridge's developing intellect. He had formed decided ideas on the pedantry of book learning while in Göttingen. "I find being learned is a mighty easy thing, compared with [?any study] else. My God! a miserable Poet must he be, & a despicable Metaphysician [?whose] acquirements have not cost more trouble & reflection than all the lea[?rning of] Tooke, Porson, & Parr united." Learning alone was just a form of idleness, and "Learning without philosophy a *Cyclops*."[70] In his first winter in the Lakes, Coleridge copied a passage from Hermes Trismegistus into a notebook, indicating just one of his motives for pursuing science and philosophy. The inseparability of science from philosophy in this passage would have been among its principal attractions for Coleridge: "And he who seeks to be pious pursues philosophy. Without philosophy, it is impossible to be truly pious; but he who has learned what things are, and how they are ordered, and by whom, and to what end, thanks the Maker, deeming him a good father and kind fosterer and faithful guardian."[71] Coleridge required that scientific theories should conform to the evidences of nature, while insisting that nature was more than a conglomeration of sense data. Reason was also in man and in nature, God's reason and man's. Scientific theories had to conform to reason as well as to sense, and the creative scientist was therefore a philosopher.

Coleridge at first believed that Davy was truly a philosophical scientist.[72] His metaphysical contributions must have come primarily from the laboratory. He enjoyed metaphysical speculations, but his response to Coleridge's urgent inquiry was to wonder whether it was possible to give sufficient consistency to his theories, "which have been constantly altering and undergoing new modification."[73] But Coleridge believed that a great scientist had to be a great philosopher. Davy's greatness in philosophy was not apparent; Coleridge concluded that it must therefore be implicit in his chemistry. When Coleridge was sufficiently distant from Davy to achieve a balanced view of his verse, he concluded likewise that his chemistry contained his true poetry. Poet, scientist, and metaphysician were all active in acquiring living ideas that were "essentially one with the germinal causes in nature." Coleridge, looking back on his earliest acquaintance with Davy, was to laud him as "the Father and Founder of philosophic Alchemy, the Man who *born* a Poet first converted Poetry into Science and *realized* what few men possessed Genius enough to *fancy*." Coleridge's criticism of Davy's verse was enthusiastic and perceptive, and Davy was wise enough to take Coleridge's advice.[74]

Through the summer and early autumn of 1800, Coleridge pressed Davy to join him, read his publications, and eagerly followed his career. He asked Davy to send him his *Researches on Nitrous Oxide*, and referred Davy to an account in Moritz's *Magazine for Experimental Psychology* (2 [1784], 12), which described symptoms similar to those produced by breathing the gas. Coleridge's concern with experimental psychology, his reference to a German source, his recollection of Davy's work in chemical physiology and of its psychological import, were typically succeeded by an account of Wordsworth's experience in which association excited sensation. Here was further material for their common interest in psychosomatic phenomena, depending equally on mental and physical science. Davy's work was clearly destined for glory. "Work hard, and if Success do not dance up like the bubbles in the Salt (with the Spirit Lamp under it) may the Devil & his Dam take Success! – . . . Davy! I *ake* for you to be with us."[75]

Davy helped to see the new sheets of the *Lyrical Ballads* through the press,[76] but was not as enthusiastic a correspondent as Coleridge, nor did he come north that autumn. Coleridge was vastly disappointed, but consoled himself with the thought that "from Davy's long silence I augured that he was doing something for me – I mean for me inclusive, as a member of the Universe." He was right. Davy had seized on Volta's invention of the pile, or battery, and had begun his brilliant train of galvanic researches, presenting the possibility of a new dynamical chemistry.[77]

Coleridge still hoped that Davy would visit him – to climb the hills, talk philosophy and chemistry, and make puns – but Davy had been busy with experiments, and ill besides. Coleridge was anxious lest Davy's chemical researches might have exposed him to unwholesome influences. "There are *few* Beings both of Hope & Performance, but few who combine the 'Are' & the 'will be' – For God's sake therefore, my dear fellow, do not rip open the Bird, that lays the golden Eggs." Personal sympathy and solicitude were still the mainsprings of Coleridge's interest in chemistry. When William Calvert, whose brother had left Wordsworth a legacy, proposed to set up a chemistry laboratory in the Lakes, and to study chemistry there with Wordsworth and Coleridge, Coleridge was enchanted. Here at last was an opportunity of learning chemistry and thus of sharing more closely in Davy's work. "Sympathize blindly with it all I do even *now*, God knows! from the very middle of my heart's heart," he wrote to Davy; "but I would fain sympathize with you in the Light of Knowledge." Calvert already had "an electrical machine and a number of little nick nacks connected with it." Coleridge now asked Davy for advice about additional apparatus and chemical texts. He proposed taking Nicholson's *Journal of Natural Philosophy* to keep abreast of Davy's work, wryly admitting that his passion for science might be "but *Davyism*! that is, I fear that I am more delighted at *your* having discovered a Fact, than at the Fact's having been discovered."[78]

"Davyism" was a part of the story. But Coleridge was also enlarging his knowledge of man as intellectual being and as animal, through science, philosophy, and psychology. He was coming to grips with the metaphysical books he had brought back from Germany, and looking with critical care at the tradition of British empiricism. John Locke was his principal target, Locke with his epistemology limited by the senses, with his fragmented, atomistic, and incoherent world of little things, and with what Coleridge clearly perceived as his plagiarisms from Descartes. "Mr Locke supposed himself an *adder* to Descartes – & so he was in the sense of *viper*."[79] Now Locke was the philosopher whom the *philosophes* of the Enlightenment had associated with Newton. They believed that Newton had produced the science, Locke the corresponding philosophy of science.[80] Coleridge's analytic scrutiny of Locke took him back more carefully to Newton. The more Coleridge looked into Newton's work, the less he thought of him. In a letter to Tom Poole that he afterward begged Poole to destroy, Coleridge confided that he believed that "the Souls of 500 Sir Isaac Newtons would go to the making up of a Shakspere or a Milton." Nevertheless, he planned to master all Newton's works within the year. He began with the *Opticks*, as being an easier work than the *Principia*:

> I am exceedingly delighted with the beauty & neatness of his experiments, & with the accuracy of his *immediate* Deductions from them – but the opinions founded on these Deductions, and indeed his whole Theory is, I am persuaded, so exceedingly superficial as without impropriety to be deemed false. Newton was a mere materialist – *Mind* in his system is always passive – a lazy Looker-on on an external World. If the mind be not *passive*, if it be indeed made in God's Image, & that too in the sublimest sense – the Image of the *Creator* – there is ground for suspicion, that any system built on the passiveness of the mind must be false, as a system.[81]

Still, the experiments *were* worth repeating. Coleridge asked Poole to bring him three prisms, and with them and metaphysics as his guide, found himself an unwilling captive to the geometrizing spirit. No longer did he see in the mountains

> The lovely shapes and sounds intelligible
> Of that eternal language, which thy God
> Utters, who from eternity doth teach
> Himself in all, and all things in himself.

He now saw instead of the motion and divine symbolism of the hills only a static series of abstract curves. This was "abstruse research," deadening the pains of sleepless nights – for opium was taking its toll – while suspending his "shaping spirit of imagination." The prisms found their use in optical experiment – Coleridge rubbed a cat's back in the dark to try refracting the sparks with a prism, only to be scratched for his pains.[82]

Davy and Newton were Coleridge's principal but not sole scientific fare at this time. He heard of William Herschel's discovery of the infrared spectrum and, mindful of Davy's work on light and colors, and perhaps also of Tom Wedgwood's, wondered whether this extension of the spectrum would revolutionize chemical theory. Coleridge's chemical studies were not going well: "As far as *words* go, I have become a formidable chemist – having got by heart a prodigious quantity of terms &c to which I attach *some* ideas – very scanty in number, I assure you, & right meagre in their individual persons. That which most discourages me in it is that I find all *power &* vital attributes to depend on modes of *arrangement* – & that Chemistry throws not even a distant rush-light glimmer upon this subject. The *reasoning* likewise is always unsatisfactory to me – I am perpetually saying – probably, there are many agents hitherto undiscovered."[83]

Coleridge was frequently ill, depressed in body and mind, discontented first with the state of British philosophy and then with the science resting on it. Newton's physics was vitiated for him. Now contemporary chemistry appeared to be equally mistaken in its foundations. Power, arrangement, organization, unifying ideas – all seemed lacking. Davy's "Essay on heat and light" was pilloried by the reviewers, much though Beddoes and Coleridge liked it.[84] But it had at least proposed a relation between the powers of matter and of light in conformity with a universal law of nature. Herschel's work explicitly, and more respectably, related heat, light, and chemical action to one another; yet it did not readily fit in with contemporary chemical theory. Beddoes and Davy had thrown out suggestions offering some hopes for a chemical reformation, but nothing had yet come of them. The imperfections of Lavoisier's Lockean chemistry were as palpable as those of Newton's "Little-ist" physics. Beddoes's magnificent hopes for pneumatic medicine – breathing the air from cowhouses to combat consumption, taking nitrous oxide as a specific cure for paralysis, and the rest – were illusory. In the winter of 1801, Coleridge confided bitterly to his notebook: "Beddoes hunting a Pig with a buttered Tail – His whole Life an outcry of *Ευρηκας* and all eureekas Lies."[85] Without Davy, Coleridge would surely have abandoned the miserable science of chemistry.

But there was Davy, newly established in the Royal Institution, where he had been made assistant lecturer and then, in June 1801, lecturer in chemistry. Davy's public lectures were such a social and financial success as to help to divert the Royal Institution from its original philanthropic aims. Davy was ecstatic at his reception, whereas Coleridge was alternately fearful that success would corrupt Davy and confident that his intellectual powers would protect his moral character. "There does not exist an instance of a *deep* metaphysician who was not led by his speculations to an austere system of morals."[86]

Davy's lectures: chemistry and metaphor

In the late autumn of 1801, Coleridge came to London, where he wrote for the *Morning Post*, pursued his metaphysical inquiries, and saw Davy – less than he hoped, for Coleridge was again wretched. After Christmas he returned with Poole from Stowey, in time for Davy's lectures in January 1802.

Davy sent him an inscribed copy of the syllabus.[87] He was in fact offering two courses, one in the morning on "General Chemistry" and the other in the evening on the "Outlines of Chemical Science, and the Chemistry of the Arts." Coleridge was particularly concerned with the morning lectures. He took notes, which do not correspond to the full span of the syllabus, but do match Davy's own summary. Coleridge probably attended the complete series of morning lectures, as he had planned.[88]

The introductory lecture stressed the central role of chemistry in science, together with chemistry's widespread utility in the operations of everyday life. The intellectually creative chemist could thus contribute to social and moral improvement. Here was reason enough for Coleridge to study chemistry, but he also stated that he attended Davy's lectures to enrich his stock of metaphors.[89]

Coleridge was to explain that metaphors provide illustrations for a theme by transferring names and descriptions from their accustomed objects to different but analogous objects. They thus suggest or even create a new structure of relations in language and thought. Coleridge was not only a poet; he was also a would-be educator, for whom one of the primary goals was the inculcation of the habit of seeking and finding relations in mind and nature. Education was essentially concerned with refining the sense of relation.[90] Coleridge, looking back to his attendance at Davy's lectures, recalled or thought he recalled how chemistry could break down accustomed artificial divisions of thought.[91] He often used chemistry and chemical processes as images for human psychology. This was possible in part because he believed that chemical changes, reactions leading to products, could themselves only be described metaphorically: "The Theorist who rejects the metaphor because metaphorical must (supposing him consistent) deny Chemistry to be Chemistry."[92] The structure of chemical metaphor – combination, exchange, saturation, affinity – embedded in the language and grammar of chemistry, reflected for him the structure of psychology, the structure of human thought and of human language. Chemical nomenclature, the assembly of conceptual chemical tools, "as far as it was borrowed from Life & Intelligence, half-metaphorically, half-mystically, may be brought back again (as when a man borrows of another a sum which the latter had really previously borrowed of him, because he is too polite to remind him of a Debt) to the use of psychology in many instances – & above all, in the philosophy of Language – which ought to be experimentative & analytic of the elements of meaning, their ~~single~~, double, triple and quadruple combinations, – of simple aggregation, or of composition by balance of opposition."[93]

The metaphor that Coleridge drew from chemistry was not mechanical. He was at great pains to distinguish the sciences from one another, kept mechanics distinct from chemistry, and used chemistry to illustrate the difference between synthesis and juxtaposition. Oil and water could merely be juxtaposed; the addition of an alkali produced a synthesis or combination. Imagination was the synthetic power in mind.[94] Chemical metaphor could thus take one beyond the psychology of association to creativity in thought and language.

Davy expressed a complementary view of the relations among thought, language, and chemical science. "[We] use Words for Ideas as we use signs for collections of units in algebra . . . if we were accurately to examine the progress of intellect we shall find that . . . the Laws of the universe have owed their origin more to the combinations of terms and propositions than to the perpetual consideration of ideas representing facts." Davy enlarged on this in his lectures, because he believed that chemistry was the science "the most capable of all others of being expressed by Language."[95]

This exegesis has run well beyond Coleridge's views of mind in 1802, and has drawn from sources ranging over more than twenty years. But even at this date, through his philosophical reading and his fascination with creativity, he assumed that there existed a harmony between the powers and forms of mind and the powers and forms of nature.[96] His poems suffice to make this clear. The subsequent development of his use of chemical metaphor to illustrate psychology and the philosophy of language shows the maturation of his psychological understanding, but shows no change in the way in which he used chemical metaphor. Yet Coleridge did in truth enrich his stock of metaphors by attending Davy's lectures.

Davy had presented his material fully, clearly, and conventionally. His fundamental organizing principle for the lectures on general chemistry was the doctrine of elective affinities developed in England and France in the eighteenth century; his nomenclature was mostly Lavoisier's; and his judicious use of British and continental literature was altogether unexceptionable. Coleridge was therefore offered a very fair statement of contemporary chemical theory, which he summarized in more than sixty pages of notes. His notes were systematic and accurate, evincing tolerable familiarity with the subject. He had, for example, no problem with the esoteric names for the various earths.

Coleridge's grasp of chemical nomenclature was precise. His observation of Davy's demonstrations was equally so, with a poet's, even a painter's, eye for color:

> Ether . . . burns bright indeed in the atmosphere, but o! how brightly whitely vividly beautiful in Oxygen gas . . . Zinc burns with a very bright flame, the lower part white, the conical part purple blue.
>
> Tin burns with a flame rather whiter than zinc – and a diffusion of a bright violet color or crimson blue . . .
>
> Copper burns with a blue flame, rather green at the Top.[97]

Davy's lectures were a delight to see as well as to hear. Coleridge rejoiced at Davy's progress, while deploring the destructive effects of worldly success. Davy, a being of hope promising most proudly, had to strangle the twin serpents of dissipation and vanity. Coleridge was caught between disillusionment and admiration and suffered disproportionately. In January 1804 he told Southey of a nightmare, in which he dreamed "that I came up into one of our Xt [Christ's] Hospital Wards, & sitting by a bed was told that it was Davy in it, who in attempts to enlighten mankind had inflicted ghastly wounds on himself, & must henceforward live bed-ridden. The image before my Eyes instead of Davy was a wretched Dwarf with only three fingers . . . I . . . burst at once into loud & vehement Weeping."[98]

Davy was his friend. Coleridge saw a good deal of him in London between January and March 1804, attended some of his lectures at the Royal Institution, discussed chemistry and natural philosophy with him, and fretted at his temptations. "Called on Davy – more & more determined to mould himself upon the age in order to make the age mould itself upon him."[99]

Opium was meanwhile bedeviling Coleridge's relations with his friends, causing physical as well as mental anguish. Coleridge still seems to have half-believed that his sufferings were relieved rather than caused by opium, and hoped vainly that a more genial climate than that of Keswick would restore him to health. At the end of January 1804 he breakfasted with Greenough, "& from him first heard & thought of Catania & Mount Ætna." John Dalton, whose work in meteorology was to lead him to his atomic theory, met Coleridge through the Royal Institution and convinced him of what he already knew, that Keswick was a singularly wet place. Surely Sicily would be better for him. But in the end he determined on Malta.[100] The greatest consolation on the eve of his departure was a much treasured letter from Davy:

> In whatever part of the world you are, you will often live with me; not as a *fleeting idea*; but as a *recollection possessed* of *creative* energy, as an *imagination winged with fire* inspiring & rejoicing. – . . .
>
> May blessings attend you my dear Friend. Do not forget me. We live for different ends & with different habits & pursuits; but our feelings with regard to each other have I believe never altered – They must continue, they can have no natural death; & I trust that they can never be destroyed by fortune, chance or accident.[101]

Their feelings, or at least Coleridge's for Davy, underwent several reverses. But there was never any slackening of Coleridge's interest in Davy's work. One series of researches in particular, on the chemical agencies of electricity, impressed Coleridge with its brilliance, and later aided him in formulating his dynamic alternative to the chemistry of Dalton and Lavoisier, exemplifying a distinct structure of science and of the philosophy of nature.

Coleridge and Davy's galvanic researches

When Coleridge and Davy first met, Davy was filled with boundless ambition: "Yet are my limbs with inward transports thrilled, and clad with new-born mightiness around."[102] But how could he exert his genius? With Beddoes's encouragement, he had already criticized Lavoisier's table of elements and challenged his theories of caloric and of respiration. Criticizing was not enough, he must be doing. He had to tackle Lavoisier's system at its weak points, topple and supplant it. In 1800, Volta's announcement of his invention of the electric pile, dramatically enlarging the arsenal of chemistry, provided Davy with the weapon he sought. Here was a seemingly unbounded source of power for overcoming affinities and revealing the constitution and arrangement of bodies. Davy, assuming Joseph Priestley's mantle, would perform with current electricity what Priestley had predicted for static electricity.[103]

The voltaic pile produced its own surprises. William Nicholson and Sir Anthony Carlisle used it to decompose water, and were perplexed to find hydrogen and oxygen evolved separately at the two electrodes. Why were the gases not evolved together? Here was a clue to the role of electricity in chemical operations, and a fertile new field for research. Davy energetically and delightedly entered this field, investigating the action of the pile as if Volta had made it for him.[104]

He inclined to the belief that the electrical effects produced by the pile were "*somehow*" brought about by chemical changes, and discovered alternative ways of constructing galvanic piles. He was at once cautious and exuberant, informing his former patron Davies Giddy that at Clifton he had met "with unexpected and unhoped-for success. Some of the new facts on this subject promise to afford instruments capable of destroying the mysterious veil which Nature has thrown over the operation and properties of ethereal fluids."[105]

Coleridge, removed to Keswick, looked eagerly for reports of Davy's work. He was "right glad, glad with a *Stagger* of the heart," to see him writing again. He read in the *Morning Post and Gazetteer* an advertisement for Davy's "*Galvanic Habitudes of Charcoal* – Upon my soul, I believe there is not a Letter in those words, round which a world of imagery does not circumvolve." Davy studied some of the physiological effects of the pile and their significance for "philosophical medicine."[106] Chemistry had already appeared as the key to problems in physiology and perception. Now galvanism, as electrochemistry was called, promised deeper knowledge. Davy confided to Coleridge in November 1800 that he had made "some important galvanic discoveries which seem to lead to the door of the temple of the mysterious god of Life." Coleridge, rejoicing, cheered him on: "Success, my dear Davy! to Galvanism & every other ism & schism that you are about."[107] Davy gave a course of lectures on galvanism at the Royal Institution early in 1801. Coleridge, learning of this, rightly assumed that Davy had made important discoveries in galvanism, "or I should be puzzled to conceive how that subject could furnish matter for more than one Lecture." He had heard of the lectures with a sensation akin to galvanic shock, and wished fervently that he were one of Davy's audience.[108]

Davy's researches went steadily on, but he was soon frustrated by the lack of dramatic discovery. He knew he was capable of greatness, which exasperatingly eluded him. His notebooks were his confidants. At Bristol he had expressed his ambition crisply enough: "Davy and Newton." Now, in 1805, he experienced "restlessness of thought, power superior even to will, ardent, but indefinite hope."[109]

In 1806 that power clearly emerged, disciplined and directed as never before, in his most brilliant and meticulous paper, the Bakerian Lecture "On some chemical agencies of electricity." The lecture was delivered in November 1806 and published in 1807. Davy had tackled the problem of electrochemical action, and had arrived at a theory of electrolytic action in which electricity functioned as a force essential to matter, and matter was particulate. He had demonstrated a relation between electricity and chemical affinity, and had shown connections between static and galvanic electricity, chemical affinity, heats of reaction, and the electrochemical series. He had also shown how Volta's pile, susceptible of indefinite enlargement and consequent increase in power, might reveal the true elements of bodies.[110]

This prediction was followed swiftly by its impressive fulfillment. In October 1807 he achieved the electrolytic decomposition of the fixed alkalis, thus discovering sodium and potassium: "Capl. Expt." Not content to rest there, he predicted the analogous decomposition of the alkaline earths, a prediction likewise fulfilled in 1808. Chemistry could be organized and prosecuted through the recognition that chemical process was governed by the interplay of positive and negative powers. Chemical affinities could be correlated if not identified with a graduated sequence of electrical powers, and chemical elements could be identified and classified in relation to an electrochemical series from the most electropositive elements, like sodium, to the most electronegative ones, like oxygen.[111]

Coleridge had complained to Davy in 1801 that in chemistry he found "all *power* of vital attributes to depend on modes of arrangement, and . . . chemistry throws not even a distant rush-light glimmer upon this subject." Davy's Bakerian Lectures made chemistry derive from power and arrangement, and promised a heartening reformation in the science. Coleridge was positively exalted by Davy's discoveries – "discoveries more intellectual, more ennobling and impowering human Nature, than Newton's!"[112] His enthusiasm was widely shared. The brilliance and significance of Davy's researches were promptly recognized. Beddoes described the first Bakerian Lecture as almost the greatest thing done in chemistry since Black's work – no mean tribute from one who regarded Black as the source of all wisdom in chemistry. J. J. Berzelius, the greatest chemist of the nineteenth century and far from over-partial to Davy, called the paper "one of the best . . . which has ever enriched the theory of chemistry." H. Brougham in the *Edinburgh Review* hailed the investigation as unsurpassed "in modern times, for closeness, copiousness, and minute accuracy." He went on, in rare harmony with Coleridge's sentiments, to remark that it was "no small proof of Mr. Davy's natural talents and strength of mind, that they have escaped unimpaired from the enervating influence of the Royal Institution; and indeed grown prodigiously in that thick medium of fashionable philosophy."[113]

Triumph was succeeded by illness. When Coleridge reached London in the mail coach from Bristol on 23 November, he found Davy seriously ill after his "March of Glory, which he has run for the last six weeks – within which time", he informed Dorothy Wordsworth,

> by the aid and application of his own great discovery, of the identity of electricity and chemical attractions, he has placed all the elements and all their inanimate combinations in the power of man; having decomposed both the Alkalies, and three of the Earths, discovered as the base of the Alkalies a new metal, the lightest, most malleable, and most inflammable substance in nature – a metal of almost etherial Levity – and which burns under water by merely being placed by the side of Sulphur. He has proved too, that by a practicable increase of electric energy all *ponderable* compounds (in opposition to *Light* & *Heat*, magnetic fluid, &c.) may be decomposed, & presented simple – & recomposed thro' an infinity of new combinations.[114]

Coleridge's account was exaggerated. The resolution of all ponderable compounds into their elements had not been proved, however likely it may have seemed in consequence of Davy's view of the electrical nature of chemical affinity. But Davy's real achievement was striking enough.

"I was told by a fellow of the *Royal Society*," Coleridge told Dorothy, "that the sensation produced last week, by the reading of his Paper [his second Bakerian Lecture] there, was more like stupor than admiration – & the more, as the whole train of these discoveries have [has?] been the result of profound Reasoning, and in no wise of lucky accident." The account concluded with opinions going decidedly beyond Davy's lecture, compounded of Davy's conviction that nature was a harmonious and fundamentally simple whole, Coleridge's belief that metaphysics served to inculcate principles of right action, and their joint belief in the unity of natural and moral law:

> Davy supposes that there is only one power in the world of the senses; which in particles acts as chemical attractions in specific masses as electricity, & on matter in general, as planetary Gravitation. Jupiter est, quodcumque vides; when this has been proved, it will then only remain to resolve this into some Law of vital Intellect – and all human Knowledge will be Science and Metaphysics the only Science. Yet after all, unless this be identified with Virtue, as the ultimate and supreme Cause and Agent, all will be a worthless Dream. For all the Tenses and all the compounds of *Scire* will do little for us, if they do not draw us closer to the Esse and Agere.[115]

For Coleridge, in the winter of 1808, Davy's achievement was the world's honor and glory. Believing that the *Edinburgh Review* had praised Davy insufficiently, he drafted a counterblast: "It is high Time," he told Davy, "that the spear of Ithuriel should touch this Toad at the ear of the Public." He revived his interest in chemistry, and once back at the Lakes exchanged chemical news with Davy. He read an account of Davy's first lecture of the winter of 1809, and though expressing reservations about the mingling of science with theosophy, nevertheless found himself overwhelmed by the proofs Davy had adduced of the analogy and moral connection between man's mind and God's. "Shame be with me in my Death-h[?our if] ever I withold or fear to pay my just debt of d[?eserved] Honor to the truly Great man, because it has bee[?n my] good lot to be his Contemporary, or my happiness [?to] have known, esteemed, and loved as well as admired him." Such convictions led Coleridge to claim that "Humphrey Davy in his Laboratory is probably doing more for the Science of Mind, than the Metaphysicians have done from Aristotle to Hartley, inclusive." But in 1812 or later, Coleridge added a comment: "Alas! Since I wrote the preceding note, H. Davy is become Sir Humphrey Davy & an *Atomist*!"[116]

Between the principal note and its later nullifying coda, Davy had been knighted, had married a very wealthy bluestocking, Jane Apreece, and, perhaps worst of all, had been convicted by Coleridge of the heinous crime of atomism. Davy had certainly adopted and been adopted by the establishment in science and society. But he had not made any real change in his views about atomism.[117] It was Coleridge who had shifted, in science and philosophy, to a position from which Davy's Newtonian use of atoms and powers now appeared as the rankest corpuscular materialism, and the science that had supported Unitarianism was inadequate for growing trinitarian convictions. Coleridge was adding to his knowledge of Kant an acquaintance with the writings of Schelling and of Steffens, finding in them genial coincidences with the development of his own thought and feelings. A more fully dynamic philosophy of a living nature made Davy's essentially eighteenth-century compromise in matter theory appear tainted by all the false principles that Coleridge had come to condemn in the Enlightenment. Coleridge had encountered *Naturphilosophie* with its new vision of the world. The old vision seemed all the bleaker and more sterile in contrast.

2

SURGEONS, CHEMISTS, AND ANIMAL CHEMISTS

COLERIDGE'S PRODUCTIVE MIDDLE YEARS FROM THE *BIOGRAPHIA LITERARIA* TO *AIDS TO REFLECTION*

German literature and scientific preoccupations

When Coleridge returned from Malta in 1806, he had left Unitarianism behind him and that autumn "was making up his mind how he could become a full Trinitarian." And his way toward trinitarian orthodoxy was "along the high metaphysical road."[1] His support along that road came in part from Plato, from the Neoplatonists, and from recent German philosophy. Because the science that best supported post-Kantian idealism was, substantially although not exclusively, German, Coleridge increasingly immersed himself in German philosophical and scientific writings. Here was his introduction to a new philosophy of nature.

In the late summer of 1809 he drew up yet another plan of education, comprising logic, mathematics, ancient history, and "at the same time Lectures on the transition of Mechanics into Chemistry, with such experiments only as are necessary to demonstrate the Numeration-table of the known Undecompounds, Elements of our present knowledge." Then came psychology, in familiar association with chemistry, followd by metaphysics, theology, and modern history.[2] The novelty in this plan compared with earlier ones is the proposal of lectures on the "transition of Mechanics into Chemistry." The phrase is susceptible of more than one interpretation, an ambiguity probably intentional and significant for Coleridge. Chemistry had been a part of physics, indeed of mechanics, in seventeenth-century Newtonian natural philosophy. In the early years of the nineteenth century, John Dalton had proposed a specifically chemical atomic theory and the nature philosophers of Germany were developing their very different chemistry of powers. Their two styles of chemistry had little in common except that in both of them the old reduction of chemistry to mechanics was no longer valid. The transition to which Coleridge referred was thus partly a historical one. Qualities of matter, formerly explained by mechanics, now required a chemical explanation, but the end of the old reductionism placed chemistry and mechanics in a new relation. The *Naturphilosophen* saw nature as the product of powers, and the sciences were identified by characteristic powers, such as magnetism and chemical affinity. What was the new relation of the sciences? What were the transitions between them? These became key questions for Coleridge. He was concerned with the relation of powers to qualities, and with the way in which powers developed.

As early as the spring of 1809 he was asking such questions about the power of life.[3] That he regarded life as a power is in itself significant. Albrecht von Haller, John Hunter, and Blumenbach were among the founders of a tradition in physiology that took a similar view of life.[4] But Coleridge's vocabulary in discussing life also suggests an early indebtedness to the *Naturphilosophen*, and especially to Schelling.[5] Blumenbach's formative power provided a suggestive metaphor that was almost a model for imagination, raising questions about the relation between imitation and imagination.[6] *Biographia Literaria*, dictated in 1815, and the *Theory of Life*, begun in or around 1815, contain partial resolutions of this complex of problems.

Immanuel Kant, Johann Gottlieb Fichte, F. W. J. von Schelling, J. N. Tetens, and Henrik Steffens furnished, after 1809 and until the early 1820s, his principal aids in framing and answering questions about the philosophy of nature.[7] Schelling – not just the early Schelling, but also the later, for whom religion and philosophy were to be one – was important for Coleridge's overall development;[8] Steffens, philosophically far less significant, was important for Coleridge's approach to science, furnishing him with a wealth of philosophically digested scientific information. Steffens was born in Stavanger, Norway, in 1773, the son of a surgeon from Holstein. After studying at the University of Copenhagen he won a travelling scholarship that took him successively to Hamburg, then to Kiel, and, in 1798 following a walking tour in the Harz, to Jena, where he studied under Schelling and became an assistant professor (*extraordinarius*) of philosophy. He next removed to Freiberg, where he met and studied under the geologist and mineralogist A. G. Werner. In 1802–4 he was back in Denmark, and then he took a chair at Halle. He was in the Prussian army in 1814; when the war ended, he became professor of natural history at Breslau, moving in 1831 to a similar chair in Berlin, where he died in 1845.[9] He combined the dynamic philosophy of Schelling with a detailed knowledge of the natural sciences, especially the nonmathematical ones.

Coleridge set about acquiring the works of the new school. In November 1813 he wrote to Henry Crabb Robinson, requesting works by Fichte and Schelling. This would "not only oblige but really serve me – for I have a plan maturing, to w[?hich] that work would be serv[?iceable]." Coleridge had already asked Robinson to send him any works bearing on the "Neuere, neueste, und allerneueste Filosofie," and at the same time had asked for Goethe's *Farbenlehre.* It seems, however, that he first set about formulating a systematic and methodical plan of studies in 1813, a year when opium addiction had brought him to an otherwise low ebb. In 1815 or 1816 he drew up a list of German books he wanted. In the summer of 1816 he proposed his letter on German literature, including the sciences, insofar as they were illuminated or influenced by philosophy.[10] September found him writing to his bookseller, Thomas Boosey, asking if he could borrow

> all the numbers, you happen to possess of the Allgemeine Literatur-Zeitung, for a few weeks – I think, I may venture to promise that the result will be of equal advantage to yourself as to me. All the Articles respecting Medicine, Chemistry, Magnetismus [zoo-magnetism and animal magnetism], and the Natur-philosophie with it's opponents in general, are my sole present Object . . . It might appear presumptuous and arrogant if I should say that *I* know no literary man but myself fitted for the conduct of a periodical Work, the Object of which should be to create an enlightened Taste for the genuine productions of German Genius in Science, and the Belles Lettres; but the arrogance would disappear on a closer view.[11]

In the following spring, Coleridge asked Boosey for a complete list of the textbooks used in German universities, and mentioned that Boosey's son had "a list of all Schelling's and Steffens's works that I already possess; and a general order to purchase whatever of these Authors is not included in that list."[12]

Coleridge was assiduous in pursuing his program of reading and thinking. Nothing as formal and regular as a fortnightly or monthly letter ever came of it, yet he was accurate rather than arrogant in claiming to be the literary man best prepared for composing such a letter. His interest was in learning how German thought could aid man's knowledge of himself and of nature. He sustained and pursued this interest with determination and critical acumen, so that his use of German sources, here as elsewhere, is more constructive than derivative.[13] His explorations of natural knowledge were at once tentative and progressive, few of them reaching any definitive conclusions. What emerges is a method of inquiry, and a new perspective on nature, rather than a finished system of natural knowledge. The schemes that Coleridge did propose were mutable, functioning heuristically rather than dogmatically. He was aware of the temptations of system-spinning and abstract metaphysics, through which thinking could become an indulgence in hubris.[14]

With personal experience of the dangers of abstract thought, Coleridge was doubly drawn to the individual facts of experience. A metaphysical dream castle could be brought down to earth by asking whether it was valid for horses or dogs. A scheme of powers could be invalidated by the results of a single chemical analysis. And a system of philosophy that failed to account for the experience of facts was at best inadequate. Of Kant's *Critique of Pure Reason*, he remarked: "The perpetual and unmoving Cloud of Darkness that hangs over the Work to my 'mind's eye', is the absence of any clear account of – Was ist Erfahrung? What do you mean by a fact, an empiric reality . . . ? I apply the categoric forms to a tree – well! But first, what is this Tree? How do I come by this Tree?" His speculative science was controlled by empiricism, and he checked his philosophy against experimental fact.[15]

Facts were one refuge from the thinking disease. Another refuge was the subordination of metaphysics to a practical goal. Coleridge had clear moral and theological goals, implicit in his plans for education, explicit in his *Lay Sermons*: in *On the Constitution of the Church and State*, with its clerisy and national church; in *The Friend*, with its emphasis on principles; in *Aids to Reflection*, with its distinction between reason and understanding; and in his *Opus Maximum*, where science and metaphysics are subordinated to an understanding of God's reason in the world.[16]

Biographia Literaria

When Coleridge dictated the *Biographia* in 1815, the manuscript he sent to the publisher was largely about his "principles in politics, religion and philosophy, and the application of the rules deduced from philosophical principles to poetry and criticism."[17] The importance of principles was a guiding theme throughout his prose works. Coleridge suggested another interpretative key. The *Biographia* contained "a sketch of the *subjective* Pole of the Dynamic Philosophy, whose results in relation to ethics and theology were to appear later in the third volume of *The Friend*;[18] his philosophy of nature constituted the objective pole of his version of dynamic philosophy.

At the beginning of the *Biographia* he recapitulated his early discussions about Erasmus Darwin's *Botanic Garden*, and his contempt for Hume, Condillac, and Voltaire, all exponents of empirical philosophy. He attacked mechanical philosophy in all its guises, attributing its prevalence to the despotism of the eye: "Under this strong sensuous influence, we are restless because invisible things are not the objects of vision; and metaphysical systems, for the most part, become popular, not for their truth, but in proportion as they attribute to causes a susceptibility of being *seen*, if only our visual organs were sufficiently powerful." Imponderable fluids and Hartley's mechanism of association were just two of the fictions arising from the despotism of the eye and from "the mistaking the *conditions* of a thing for its *causes* and *essence*; and the process, by which we arrive at the knowledge of a faculty, for the faculty itself." Association presented the mind as essentially mechanical and passive, whereas creative minds were not merely receptive but also active. Wordsworth, for example, combined "the fine balance of truth in observing, with the imaginative faculty in modifying the objects observed." Thinking involved activity as well as relative passivity.[19]

Coleridge illustrated his argument with a detailed piece of observation: "Most of my readers will have observed a small water-insect on the surface of rivulets, which throws a cinque-spotted shadow fringed with prismatic colours on the sunny bottom of the brook; and will have noticed, how the little animal *wins* its way up against the stream, by alternate pulses of active and passive motion, now resisting the current, and now yielding to it in order to gather strength and a momentary *fulcrum* for a further propulsion. This is no unapt emblem of the mind's self-experience in the art of thinking."[20] Now if the mind was active, modifying what it perceived, it was proper to seek canons of criticism deduced from the nature of man.[21] Here was one major potential field for the application of Coleridge's principles. The modes of activity and receptivity of the mind likewise determined the requirements of clear thinking and the arts of memory, which included "sound logic, as the habitual subordination of the individual to the species, and of the species to the genus; philosopical knowledge of facts under the relation of cause and effect; [and] a chearful and communicative temper disposing us to notice the similarities and contrasts of things, that we may be able to illustrate the one by the other."[22] Coleridge, as we shall see, presents these "arts" as most clearly applicable to the sciences, although needed for all constructive intellectual endeavor.

These are important arguments. They show the inadequacy for Coleridge of the empiricist tradition in philosophy from Locke to Hartley. They also show the corresponding inadequacy of systems of extreme idealism that, like those of Leibniz and Descartes, propose an external world corresponding to yet divorced from our mental images. Coleridge's frustration was relieved by rereading Jacob Boehme, who helped him "to keep alive the *heart* in the *head*." The works of Kant meanwhile both stimulated and disciplined his thought. From Kant, whom he continued to read all his life, Coleridge moved to Fichte and thence to Schelling; in the latter's *Naturphilosophie* and transcendental idealism he found "a genial coincidence with much that I had toiled out for myself, and a powerful assistance in what I had yet to do." The fundamental problems remaining, indeed "all the difficulties which the human mind can propose for solution," were included in "the general notions of matter, spirit, soul, body, action, passiveness, time, space, cause and effect, consciousness, perception, memory and habit."[23]

Coleridge's wide-ranging explorations in science and metascience were conducted within the framework of this inquiry. He, like Schelling, discussed productivity and stressed the role of self-consciousness in knowledge: "All knowledge rests on the coincidence of an object with a subject." Here were concerns vital to him as a poet and educationist. Coleridge also shared Schelling's formulation of the goals of natural philosophy, whose highest perfection "would consist in the perfect spiritualization of all the laws of nature into laws of intuition and intellect." Natural science would end with nature as intelligence, and would thus become natural philosophy.[24]

Coleridge's account of the relation of man and mind to nature was original in its development and feeling, although derivative in its expression, furnishing him with an explanation of perception as constantly creative rather than passively receptive. The poet's special genius appeared in his idealization and unification of material furnished by perception and fancy. The symbolism of language enabled the imaginatively creative poet to mold images of nature into an ideal form. But science also, through the creative minds of men of genius, perceived and created ideas and unity in the world of nature. The *Biographia*'s philosophical and critical apparatus had provided him with the means of demonstrating complementarity between poetic and scientific genius and imagination.[25]

Great poetry, like science, required steady and sympathetic faithfulness to nature, as Coleridge argued in discussing Shakespeare, his constant exemplar of high poetic genius. Shakespeare, he asserted, had "that sublime faculty by which a great mind becomes that, on which it meditates. To this must be added that affectionate love of nature and natural objects, without which no man could have observed so steadily, or painted so truly and passionately, the very minutest beauties of the external world." Fidelity to nature was, however, not enough; great poetry also required independence from nature. Shakespeare was "no mere child of nature; . . . no passive vehicle of inspiration . . . [he] first studied patiently, meditated deeply, understood minutely, till knowledge, become habitual and intuitive, wedded itself to his habitual feelings, and at length gave birth" to his stupendous and unique power.[26]

Coleridge held himself to the same standard. The *Biographia*, together with his lectures on drama, demonstrate once again the active unity of Coleridge's mind, connecting his seemingly twice-abstracted speculations about science with the immediacy and precision of his poetic observation. He sought to apply the same principles to the creativity of scientific and poetic genius. The latter application constitutes the substance of the remainder of the *Biographia*, and is not central to this study. The former application is central in the *Theory of Life*, which he began to draft shortly after completing the *Biographia.*

The composition of the *Essay on Scrofula* and the *Theory of Life*

The *Theory of Life* was written partly in response to a contemporary debate in physiology and medicine, and partly out of Coleridge's sympathetic concern with the medical studies of James Gillman, the surgeon with whom he resided after 1816. The original manuscript has not been found; the posthumous edition of 1848 bears Coleridge's name on the title page, but states in a postscript that the editor, Seth Watson, had come to believe "that the work might with more propriety be considered as the joint production of Mr. COLERIDGE and the late Mr. JAMES GILLMAN, of Highgate." Sara Coleridge was properly indignant. The prose style is Coleridge's. The works referred to or drawn upon in the course of the book were all works that Coleridge knew, for example, Steffens's *Beyträge zur innern Naturgeschichte der Erde* (Freyberg, 1801), whereas Gillman was ignorant of German sources. The use of these works is reminiscent of Coleridge's procedures in the *Biographia*. The position stated and developed in the *Theory of Life* is congruent with Coleridge's philosophy of nature and is stated and developed in his notebooks. Finally, fundamentally similar formulations of the theory occur in letters and notes probing the problem of the nature of life from as early as 1801. The *Theory of Life* is Coleridge's own work.[27]

But if Coleridge was its author, Gillman's interests lay behind it. The *Theory of Life* is a complete and self-contained statement. It is also, at the same time, a foundation for and a sequel to the *Essay on Scrofula*, being Coleridge's attempt at satisfying the requirement stated at the end of the *Essay*: "We presume then . . . that scrofula is a constitutional disease, by which, if we attach any distinct sense ⟨meaning⟩ to our words, we must mean a derangement of some one or all of the primary powers ⟨in which the Life, and⟩ in the harmony or balance of which the health ⟨life⟩ of the human being consists. But this again involves the necessity of a distinct conception of life itself."[28] The *Theory of Life* offers such a conception.

The surviving manuscript of the *Essay on Scrofula* was transcribed from the original manuscript by Sara Coleridge, by her daughter Edith, and by another scribe, and shows where Coleridge's own revisions were. Sara noted at the end of her transcription that the last line of the original document appeared to be in Gillman's hand. Certainly Gillman had prompted the composition, but he was not its author.

In 1811, Gillman had won the Jacksonian Prize of the Royal College of Surgeons of England for his *Dissertation on the Bite of a Rabid Animal* . . . (London, 1812). In 1816, when Coleridge moved in with him, the subjects for the competition were scrofula and syphilis.[29] Gillman decided to try for the prize, and he and Coleridge agreed to work together. Gillman seems not to have written his contribution; the *Essay on Scrofula* that survives is wholly Coleridge's in method, balance, and style.[30] The *Theory of Life*, complementing the *Essay* and written after it, was equally Coleridge's. The evidence of his letters and notebooks and the date of the competition suggest that the *Essay* and the *Theory* were conceived toward the end of 1816. The questions raised in the *Theory* were to shape Coleridge's scientific enquiries for a decade, and the dates of the conception and composition of the *Theory* are therefore significant.

It is easier to be confident about the date of conception than about that of completion. By the end of 1815, Coleridge had read the works of Schelling and Steffens that he used in writing the *Theory of Life*, and in 1816 he was adequately informed about the debate between John Abernethy and William Lawrence concerning the nature of life,[31] so that he could have completed the work in 1816–17.

A different possibility is suggested by Coleridge's acquaintance with the surgeon Joseph Henry Green, with whom he worked closely on his philosophical system and on physiological issues. Green was a member of Coleridge's Thursday class and one of his keenest disciples. He became Coleridge's literary executor, and in 1865 after years of labor published two volumes of *Spiritual Philosophy: Founded on the Teaching of the Late Samuel Taylor Coleridge.* Coleridge visited Green more than once in 1817 to meet Ludwig Tieck.[32] Tieck's visit to London may have been the occasion of Coleridge's first meeting Green, but the latter had obtained the diploma of the Royal College of Surgeons in December 1815, and had set up a surgical practice in Lincoln's Inn Fields. Gillman, himself a fellow surgeon, could thus have brought them together during 1816. In 1817, on Tieck's recommendation, Green went to study philosophy in Berlin – an unusual course of action for a recently qualified English surgeon, but one attractive to Coleridge.

In 1824, Green became professor of anatomy at the Royal College of Surgeons, giving four annual courses on comparative anatomy. His lectures included discussions of the ascent of life and of the idea of life, themes explored by Coleridge in his *Theory of Life.*[33] In 1840, Green gave the Hunterian Oration before the Royal College of Surgeons, entitled *Vital Dynamics,* which examined the concept of individualization, the ascent of life through integration and totality in each part, and other concepts and phrases that are matched in the *Theory of Life.*[34] *Vital Dynamics* reiterates the basis of Green's earlier lectures at the college; it also criticizes Lavoisier's chemistry, condemns the domination of the senses, advocates a dynamic philosophy, inquires into principles, and explicitly acknowledges and quotes Coleridge. Whether through conversation with his mentor or through familiarity with the manuscript, Green knew of the substance of Coleridge's arguments in the *Theory of Life,* although he did not know of the existence of the work until he saw it at the press.[35]

Now the *Theory* was published as a self-contained essay, distinct from the *Essay on Scrofula.* The emphasis at the end of the *Essay* upon the problem of life is primarily conveyed through revisions.[36] This separation of the *Essay* from the *Theory* is in contrast to the integrated proposal for an essay that Coleridge sketched in a notebook entry of 1817, where the discussions of the law of life and of the nature of scrofula are wholly interdependent.[37]

It is possible that discussions with Green resulted in a revision of the original manuscript of the *Theory of Life*. This revision may have accompanied Green's preparation for his lectures of 1824. Coleridge drew heavily on Steffens's *Beyträge* for the *Theory*, and in the summer of 1824, Coleridge discussed the *Beyträge* in a letter to Green in which he also referred to their discussion of life, individuation, and genetic development in nature. Shortly before this, he had written enthusiastically to a friend about Green's lectures on life. Between the first draft of the *Theory of Life* and Green's lectures of 1824, Lawrence had effectively, albeit anonymously, reopened the debate by his article on life in Rees's *Cyclopaedia*. The *Theory of Life* falls into two principal parts, one on definitions of life, the other on the physiology of life. The first part fits well into the proposal in the notebook entry of 1817; the second part could have been written at the same time, but is closer to the field of Green's lectures of 1824.[38]

The notebooks show that Coleridge returned to the life sciences and to the theory of life in the early 1820s after a thorough probing of cosmology and of chemistry.[39] Coleridge discussed the theory of life with his philosophy class in or around 1823. The account of this discussion in *Fraser's Magazine* suggests that he was at that time considering the substance and looking forward to the publication of the *Theory of Life*.[40] It would thus appear that Coleridge constructed his argument about the definition of life in the first period, in the context of the debate between Abernethy and Lawrence and of Gillman's interest in the Jacksonian Prize. He probably drafted the entire work at this time. The section dealing with the physiology of life would then have been revised a few years later.

The *Essay on Scrofula:* Lawrence, Abernethy, and the definition of life

The *Essay on Scrofula* is a document of much intrinsic interest[41] but of more limited significance for Coleridge's developing awareness of science. The essay was to have been divided into four sections. Coleridge wrote the first two, presenting a history of the opinions about the nature and origin of the disease and an assessment of those opinions; Gillman's sections were to have discussed the pathology of the disease and its treatment. Coleridge's exposition was methodical, each step furnishing the basis of the next, and the criticism of old opinions was essentially logical: "It is evident that every philosophical distinction and classification of its symptoms not grounded merely on external appearances must presuppose a theory of the disease and that the propriety of the classification must depend on the justness of the theory."[42] Coleridge presented theories described in Daniel Sennertus, *Operum* (probably the edition in 3 vols., 1860); Fabricius Aquapendente, *Opera Chirurgica* (1723); Richard Wiseman, *Eight Chirurgical Treatises* (4th ed., 1705); and John Burns, *Dissertation on Inflammation* (2 vols., 1800).[43] He went on to object "to the greater number of these theories (at all events as far as ~~necessary~~ ⟨accuracy⟩ of language is concerned) that what is supposed to *produce* the disease is, if it exist at all, a *part* of it, and entitled to the name of *cause* only by priority of action, either as the earliest or the most dangerous symptom." The manuscript concludes with the argument that an understanding of the disease required an understanding of health and of life itself, "or, as it has been the fashion of late years to name it, of the living principle, though contrary to the Canons of sound philosophic research, inasmuch as it asserts by implication the truth of an *improved* hypothesis in which the fancy has been made more active than the reason."[44]

William Lawrence was the most prominent exponent in England of the doctrine of the living principle to which Coleridge took exception. John Abernethy, who like Lawrence gave lectures at the Royal College of Surgeons, was Lawrence's principal target and opponent in the debate about the nature of life. The *Essay on Scrofula* includes a tribute to Abernethy; the *Theory of Life* devotes some space to attacking Lawrence.[45] The debate was philosophical as well as physiological, and it is noteworthy that Abernethy attended Coleridge's philosophical lectures in 1818–19. There were also political overtones to the debate, because Lawrence's views became associated with atheism and French materialism in the aftermath of the Napoleonic wars. Coleridge attended Abernethy's Hunterian Oration on February 15, 1819, when Abernethy quoted Coleridge's statement that "there can be no sincere cosmopolitan who is not also a patriot."[46] After these preliminary indications of the dimensions of the debate, it is time to turn to its substance. This account will be selective, concentrating on those aspects that concerned Coleridge in framing his own defintion of life.

Abernethy and Lawrence, as the heirs and interpreters of John Hunter, both addressed the Royal College of Surgeons. Their problem was to account for the differences between living and nonliving bodies. Living bodies maintain their chemical economy and organization, obtain nutrition from their environment, and are generally capable of reproducing their kind; the same bodies after death are broken down chemically through oxidation by the atmosphere and lose their organization. Hunter had sought to identify the single essential characteristic common to all living bodies. Abernethy and Lawrence pursued this inquiry. That they differed from one another and from Coleridge in their development of the theory of life is understandable, because Hunter's expression of what theory he may have possessed was muted in his publications and subordinated to the arrangement of the physiological and pathological specimens in his collection. The arrangement of the museum illustrated the ascent of life from a number of complementary perspectives. The principal divisions of the collection included the means of producing motion; the means of increase and support – blood, digestive organs, the heart; the brain; the organ of sense; exterior coverings – skin, feathers, hair; "peculiarities" such as horns, stings, and air bladders; and the organs of generation.[47] Coleridge saw the Hunterian Collection in the Royal College of Surgeons as a splendid example of scientific method.[48] But Abernethy and Lawrence went primarily to Hunter's writings.[49] Hunter saw blood as the simplest body endowed with life.[50] Coleridge, in a marginal note to the works of Boehme, exaggerated Hunter's statement: "Now the Blood is the Life, is affirmed by Moses – and has been forcibly maintained by John Hunter." But Hunter did insist that the blood *had* life. He found life not just in solid organized bodies but also in fluid blood, and concluded that life was independent of organization: "Mere organization can do nothing, even in mechanics, it must still have something corresponding to a living principle; namely, some power."[51]

Hunter's choice of words was surely deliberate in its breadth; "principle" was one of the most appallingly multivalent words used in the eighteenth century. It could be a source of action, a fundamental element, a native tendency, a power, a law of nature, a species of matter, and many other things besides.[52] To say that life was due to a living principle might mean that it was the result of the action of a law, or of a force, or of a specific imponderable substance. Materialists and their opponents could both invoke Hunter's authority for their views, and did so. There was, however, one scientific usage in which the term "principles" had a relatively unambiguous meaning, namely, the nonmaterial sources of the different kinds of activity manifested by matter. Newton's physics provided the authoritative model for this mode of explanation. His third rule of reasoning in philosophy, the discussion of gravitation in the general scholium to the *Principia*, and the thirty-first query to the *Opticks* suggested among them that active principles, unlike the passive qualities of bodies – inertia, mobility, and the rest – were superinduced on bodies rather than essential to them, and were diverse in kind. Each newly recognized mode of activity, including gravitational attraction, could thus be explained by a corresponding principle. In the course of the eighteenth century, active principles, sometimes embodied in imponderable fluids, were subsumed as the source of gravitational, chemical, electrical, and physiological activity. Hunter's appeal to a living principle, although perhaps wisely susceptible of several interpretations, was within the Newtonian tradition.[53]

Abernethy, addressing the Royal College of Surgeons in 1814, presented Hunter's theory accordingly. He began by distinguishing between hypothesis, rational conjecture based on an incomplete series of facts, and theory, rational explanation of the cause or connection of an apparently full or sufficient series of facts. A satisfactory theory should offer a rational causal explanation. He asserted that because in the "great chain of living beings, we find life connected with a vast variety of organization, yet exercising the same functions in each," life could not depend on organization. Then came an examination of Hunter's theory, predicated upon the assumption of the passivity of matter and the need for a supervenient active principle. So far, this was close to Hunter and to orthodoxy in British physiology.[54]

Abernethy swiftly took his argument further by developing an analogy between the electrical and the vital principle. Davy's electrochemical experiments had persuaded Abernethy that electricity produced all the chemical changes observed in inanimate objects; analogy suggested that electricity also performed chemical operations in living bodies.[55]

Abernethy, sanctioned by Newton's authority and Hunter's supposed example, proposed to explain life by a material principle akin to an imponderable electrical fluid. Coleridge, as we shall see, was unhappy at this materialism. The *Edinburgh Review* interpreted Newton very differently and attacked Abernethy for his "bad arguments, in defense of one of the most untenable speculations in physiology; interspersed with not a little bombast about genius, and electricity, and Sir Isaac Newton." Joseph Adams, who was to introduce Coleridge to Gillman, promptly countered with a defense of Abernethy, and presented Coleridge with a copy of his book.[56]

Then William Lawrence stepped into the debate with his lectures to the Royal College of Surgeons. He had been Abernethy's apprentice and then his demonstrator in anatomy, and in 1806 had won the Jacksonian Prize. He paid tribute to Abernethy and Hunter, and also to X. Bichat, who defined life as the sum of the functions by which death was resisted. Lawrence devoted his second lecture to the problem of life, and followed Bichat as emphatically as he rejected Abernethy. He denied the analogy that Abernethy had invoked between vital phenomena and those studied in the physical sciences; he rejected the concept of a subtle invisible animating matter and found no evidence for an independent living principle superadded to structure.[57]

Abernethy hit back in 1817, defending Hunter, dismissing Lawrence's views on life – "they wish me to consider life to be nothing – and rejecting Lawrence's attempt to separate the sciences from one another and from religion. Modern skeptics, "by repeatedly thinking that there may be nothing which is not an object of sense, . . . at last bring themselves to believe that there is nothing, which is a positive opinion, and also a creed found to have various conveniences."[58] It was now Lawrence's turn to be exasperated. "It is alleged," he told his audience in 1817, "that there is a party of modern sceptics, co-operating in the diffusion of these noxious opinions with a no less terrible band of French physiologists, for the purpose of demoralizing mankind! Such is the general tenour of the accusation." He deplored the incursion of politics into science. He argued that physiological knowledge was gained through the senses about the senses. Physiologically speaking, Lawrence saw life as immediately dependent on organization; and he tried once more to keep physiology distinct from religion: "The foundations of morality undermined, and religion endangered by a little discussion, and a little ridicule of the electro-chemical hypothesis of life!"[59]

Lawrence was thus allied with the French in physiology, after the Terror and the Napoleonic wars had quite overturned most English sympathies with French thought. In philosophy, Lawrence accepted Hume's account of causation as constant conjunction.[60] He saw the mind as the product of matter and the growth of the senses and sought knowledge through the senses rather than through reason; and he saw no connection between science and religion. To Coleridge, these things taken together constituted a damning indictment. Abernethy in contrast stressed causal explanation and the role of reason in science, invoked the analogy of nature, and explained life by appealing to an active principle that was neither subordinate to matter nor derived from French science and philosophy. In this controversy among surgeons, Coleridge, although an outsider, would strongly support Abernethy against Lawrence. The materialism of Abernethy's vital principle was merely an unfortunate lapse.

Coleridge began his *Theory of Life* with a bow to Hunter's bust in his museum at the Royal College of Surgeons and an acknowledgment of those "profound ideas concerning Life" presented in the Hunterian collection. He believed that Hunter had possessed "the true idea of Life," but that his expression was obscured by the language and theories of the mechanical eighteenth century. Coleridge saw Abernethy as a disciple of Hunter more consistent than his master, yet even he had failed to identify the principle of life. Coleridge would seek to achieve this identification.[61]

He first attacked the definitions of life proffered by Lawrence and Bichat. Their logic was valueless: "The physiologist has luminously explained Y plus X by informing us that it is a somewhat that is the antithesis of Y minus X; and if we ask, what then is Y − X? the answer is, the antithesis of Y + X, – a reciprocation of great service."[62] A second class of definition took "one particular function of Life common to all living objects, – nutrition, for instance"; Lawrence had argued that all organized bodies "are produced by *generation*, . . . grow by *nutrition*, and . . . end by death," and the *Edinburgh Review* had presented nutrition as the characteristic function of life. Here again was an error in logic, "the assumption of causation from mere co-existence . . . ; and this, too, in its very worst form . . . *cum hoc, et plurimis aliis, ergo propter hoc*!"[63] So far Coleridge had found Lawrence and his supporters poor logicians. He went on to complain of the inadequacy of their method. Coleridge required far more from a definition than Lawrence did:

> A . . . real definition . . . must consist neither in any single property or function of the thing to be defined, nor yet in all collectively, which latter, indeed, would be a history, not a definition. It must consist, therefore, in the *law* of the thing, or in such an *idea* of it, as, being admitted, all the properties and functions are admitted by implication. It must likewise be so far *causal*, that a full insight having been obtained of the law, we derive from it a progressive insight into the necessity and *generation* of the phenomena of which it is the law . . . For it is the essence of a scientific definition to be causative, . . . by announcing the law of action in the particular case, in subordination to the common law of which all the phenomena are modifications or results.[64]

This statement, owing much to Kant, is important for an understanding of Coleridge's conception of science. It underlines the gap between Lawrence's Humean view of explanation and Coleridge's essentially Kantian one.[65]

The next section of the argument examines the historical and national context of the debate. It owes much in its detail to Steffens's *Beyträge* and effectively places Lawrence and Bichat firmly in the traditions of Enlightenment materialism, which Coleridge rejected. He also rejected the identification of vital and electrochemical powers – Abernethy did not quite make that identification, but Lawrence suggested that he did. Coleridge rejected organization as the cause of life; he would not make the house its architect and builder. And he rejected the interpretation of Hunter's vital principle as an imponderable fluid or as a physicochemical force.[66] In his *Philosophical Lectures* of 1819 he summarized his argument. Organization "must not only be an arrangement of parts together, as means to an end, but it must be such an interdependence of parts, each of which in its turn being means to an end, as arises from within. The moment a man dies, we can scarcely say he remains organized in the proper sense. The powers of chemistry are beginning to show us that no force, not even mechanical [power, can *make* life]." Coleridge thus wholly condemned Lawrence's enterprise, while guarding against Abernethy's excesses.[67]

Coleridge quoted extensively from Lawrence's objections to Abernethy's views, especially where they concerned the use of analogy. Lawrence saw electricity and life as incommensurable and thus without analogy, and he complained that Abernethy indulged in a welter of analogies: "To make the matter more intelligible, this vital principle is compared to magnetism, to electricity, and to galvanism; or it is roundly stated to be oxygen. 'Tis like a camel, or like a whale, or like what you please."[68] Coleridge pounced on this, in such a way as to shed light on his views about the nature of analogy and the relation of powers and sciences. First, life was not *like* magnetism, or *like* electricity; the difference between them was an essential part of his system, as he believed it was of Abernethy's. Lawrence discussed analogy as if it meant resemblance. Coleridge was eager to correct him:

> Analogy implies a difference in sort, and not merely in degree; and it is the sameness of the end, with the difference of the means, which constitutes analogy. No one would say the lungs of a man were analogous to the lungs of a monkey, but any one might say that the gills of fish and the spiracula of insects are analogous to lungs. Now if there be any philosophers who have asserted that electricity as electricity is the *same* as Life, for that reason they cannot be *analogous* to each other; and as no man in his senses, philosopher or not, is capable of imagining that the lightning which destroys a sheep, was a means to the same end with the principle of its organization; for this reason, too, the two powers cannot be represented as analogous.[69]

Coleridge in the *Theory of Life* was as admiring of Abernethy as he was critical of Lawrence. In July 1819 the *Quarterly Review* examined the debate; its verdict was, like Coleridge's, for Abernethy and against Lawrence, skeptic, materialist, and Francophile.[70] The review may have prompted Coleridge to return critically to the debate. Notebook entries in the summer of 1819 show that he was reexamining the problem of the relation of life to fluidity and organization, and was reading or rereading Abernethy's *Physiological Lectures* of 1817 and his *Enquiry* of 1814.[71] Not only would he not consider Lawrence's view that "Mind is the result of Structure"; he also rejected Abernethy's view that, life being given, organization accounted for the superiority of man to the animals. The moral and religious consequences of such a view were unacceptable.[72] Besides, "he seems to have forgotten, that according to John Hunter and himself Organization is itself an effect of the vis vitae." Coleridge was also highly critical of Abernethy's apparent identification of the principle of life as an imponderable fluid. "More than a year ago," he wrote in January 1818, "I endeavoured to insinuate into Mr. Abernethy (for to attempt more than to *insinuate* would be to secure a repulse, with *Abernethy*) that as long as he clung to the phantom of a *supervenience* [i.e., life added to matter], instead of evolution [i.e., life developing], and of a supervenient *Fluid*, i.e. solved Phaenomena by Phaenomena that immediately become part of the Problem to be solved – so long he would lay himself bare to the attacks of Lawrence, and the Materialists."[73]

Mutual regard prevented such criticism from being taken amiss. Coleridge and Abernethy remained on good terms.

The Society for Animal Chemistry

In *The Friend* of 1818, Coleridge eulogized John Hunter for the physiological genius and scientific method implicit in the arrangement of his museum. He went on to exhibit Abernethy, Hunter's disciple, and the eminent surgeon, Everard Home, Hunter's pupil and brother-in-law, as recent contributors to physiology through their perceptive use of right method.[74] The chemist Charles Hatchett was also included in this tribute. Abernethy was interested in the analogy between chemistry and physiology. Hatchett and Home were concerned with investigating the relations between these sciences, and the role of chemistry in physiology. They were both members of the informal but influential Society for Animal Chemistry.

The chemical behavior of living systems differed from that of dead ones. Dead bodies decomposed by combining with atmospheric oxygen, living bodies resisted oxidation. Living bodies absorbed and modified nutrient substances, but the concept of nutrition or assimilation did not apply to dead bodies. The chemical behavior of living bodies was extraordinary, and correspondingly interesting. It was also disconcertingly complex, and only in the first decades of the nineteenth century did the techniques of analytical chemistry offer sufficient precision for chemistry to have any value for the study of most physiological processes. The problems of respiration and animal heat had provided one of the few fields in which chemistry could aid physiology in the late eighteenth century. But while Black, Lavoisier, Priestley, and others were pursuing tolerably accurate quantitative studies of respiration, the analysis of relatively simple organic compounds remained crude. An error of about 20 percent by weight was not unusual in estimating the principal constituents of an organic substance in Lavoisier's day, and elements present only in small quantitites were frequently overlooked. Organic compounds, consisting principally of carbon, hydrogen, nitrogen, and oxygen, were often so closely similar in chemical and physical properties as to be hard to separate, so that samples were often impure. Results were correspondingly unreliable. Between 1789, when Lavoisier's *Traité élémentaire de chimie* was published, and 1813, when Davy published his *Elements of Agricultural Chemistry*, the analysis of organic substances became more precise and more sophisticated. In 1814 and 1815, Berzelius was publishing impressively accurate analyses of organic compounds, and techniques of separation had improved so that gum, starch, gluten, tannin, albumen, and other substances could be characterized even if they defied analysis. It was clear by the second decade of the nineteenth century that chemistry could indeed aid physiology, and equally clear that it could not explain life.[75]

The Society for Animal Chemistry was founded in London in 1812 by a group of chemists and medical practitioners who looked to enlarge physiological knowledge by combining their expertise in chemistry and the life sciences. Hatchett was perpetual president; William Thomas Brande, who lectured on chemistry at the Royal Institution beginning in 1812 and later edited the *Quarterly Journal of Literature, Science, and the Arts*, was secretary. Humphry Davy, Benjamin Brodie, who became sergeant-surgeon to the queen, and John George Children, Davy's friend and fellow enquirer in electrochemistry, were the other founding members. Davy and Home were trustees of the Hunterian collection. The Royal College of Surgeons and the Royal Institution were thus the two poles of the society. They were also the scientific institutions with which Coleridge was most closely associated. Institutional, personal, and intellectual concerns therefore worked together to make the society's activities of particular interest to him. As we shall see, he read the members' books and papers, and his study of chemistry in 1819–20, pursued as a prelude to his return to the life sciences, was closely based upon Brande's *Manual of Chemistry* of 1819. The society lasted until 1825, when Hatchett resigned the chair and Home and Brande withdrew.[76] It is tempting to speculate on possible connections between the life of the society and Coleridge's fascination with problems at the interface between chemistry and physiology.

Relations between the sciences were always at the focus of his interests. When he wrote to Gillman in 1816 with a sketch of his theory of life, he explained the importance of relations:

> In all things alike, great and small, you must seek the *reality* not in any imaginary *elements*, (ex.gr. Sodeum + Oxygen = Soda: or Soda − Oxygen = Sodeum) all of which considered as other than elementary relations (such as are the notions, *great* and *small, tall* and *short*) are mere fictions . . . The Alphabet of Physics no less than of Metaphysics, of Physiology no less than of Psychology is an alphabet of *Relations*, in which N is N only because M is M and O, O. The *reality* of all alike is . . . the *Identity* of [Alpha and Omega], which can become an object of *Consciousness* or *Thought*, even as all the powers of the material world can become objects of *Perception*, only as two Poles or Counterpoints of the same Line.

Everything in the universe existed in relation to everything else as a totality, and also in relation to its own limited system. "Now the necessity, by which L must appear as M, should M appear as N, seen in it's relation to the center of that particular System, is the *Power or Law* of such a Phaenomenon & it's Changes: and this Power seen in it's connection with Truths that exist only in the eternal Reason, & neither have or can have any adequate Exponent in the World of Particulars or Phaenomena, is the *Idea* of the POWER: and Philosophy is the Science of IDEAS –: Science the Knowledge of *Powers*."[77] This argument was intended to help Gillman understand what Coleridge meant when he talked of life as a power, and to aid his appreciation of Coleridge's development of the connection of "Physics with Physiology, the connection of matter with organization, and of organization with Life." The *Theory of Life* was the immediate object in this letter. But the conceptual apparatus – powers, laws, and relations of the sciences – adumbrated Coleridge's discussions in *The Friend* and in numerous notebook entries.

The Friend and *Aids to Reflection*

The Friend had first appeared at what were meant to be weekly intervals in 1809–10. It was revised and enlarged to a three-volume set completed in the summer of 1817 and published in 1818. The ostensible topics were politics, morals, and religion, in relation to principles.[78] As Coleridge had already shown in the *Statesman's Manual*, such topics were bound up with his view of scientific method and were illustrated by the history of science. His concept of scientific method is encapsulated in eight chapters or essays in the third volume, where it is freer from editorial emendations than in his introduction to the *Encyclopaedia Metropolitana*.[79] Even without the striking prominence of these essays in a work scarcely devoted to science, the extent to which scientific thought and information had penetrated into the texture of Coleridge's discourse is apparent throughout *The Friend.*

Coleridge was constantly preoccupied with the role of the clerisy in the nation, the responsibilities of the upper classes, and the complementary roles of the national church and state in good governance. In *The Friend*, he discussed the relations of law and religion. He recommended "that as far as human practice can realize the sharp and exclusive *proprieties* of Science, Law and Religion should be kept distinct." Well and good. But from this practical recommendation he moved directly to metaphysics stemming from a philosophy of nature. Law and science should be kept distinct because they were opposites. "THERE IS, strictly speaking, NO PROPER OPPOSITION BUT BETWEEN THE TWO POLAR FORCES OF ONE AND THE SAME POWER." In a footnote, Coleridge gave an exegesis of this proposition that began with the law of polarity, moved to its chemical illustration, and contrasted it with the absurdity of mechanism and atomism.[80] This form of argument was frequently repeated, as when Coleridge remarked that "to *fill* a station is to exclude or repel others, – and this is not less the definition of moral, than of material, *solidity*."[81] Principles were truly universal in their applicability, and could be the more convincingly applied if they were founded in the philosophy of science. Concepts were more persuasive when endowed with scientific precision. This perspective on knowledge would have appealed to a *philosophe*, were it not for the absolute primacy that Coleridge accorded to religion.

Coleridge's religion was by now full trinitarian Christianity. He would indeed have rejected this phrase as tautological, because he believed that the idea of God entailed that of a "Tri-Unity." The polar logic, in which opposites yielded a dynamic synthesis, provided a trinitarian key to metaphysics, and thus opened the way to a philosophy of science and a metascience in harmony with Coleridge's religious convictions. This harmony was necessary, because he believed that reason in all its modes had to be self-consistent. He saw Christianity as distinguished by the unique frequency with which its scriptures commanded mankind to seek after knowledge as a sacred duty.[82] Coleridge interpreted this as an injunction to pursue the knowledge of self and of nature. He revolted at the notion of natural theology furnishing evidences of Christianity,[83] but saw no incompatibility between the study of science and the practice of religion. His distinction among sense, reason, and understanding reinforced this view.[84] Perception came through the organs of sense; understanding organized perceptions into experience; reason subordinated sense and understanding to "ABSOLUTE PRINCIPLES OR necessary LAWS: and thus concerning objects, which our experience has proved to have *real* existence, it demonstrates moreover, in what way they are *possible*, and in doing this constitutes *Science*." This was the secondary sense of reason. Coleridge also gave it a primary spiritual sense, as "an organ identical with its appropriate objects. Thus God, the Soul, eternal Truth, &c. are the objects of Reason; but they are themselves *reason*. We name God the Supreme Reason; and Milton says 'Whence the Soul *Reason* receives, and Reason is her Being.' Whatever is conscious *Self*-knowledge is Reason."[85]

Coleridge's distinction between primary and secondary forms of reason, analogous to his distinction between the primary and secondary imagination, showed both the difference in degree between scientific and religious knowledge and the parallels between them. Science had a lower but still real dignity. Religion and science both conduced in differing degrees to man's greater knowledge of himself in relation to God and to God's creation. Both revealed true principles; and true principles were "most important and sublime Truths . . . Thus the dignity of Human Nature will be secured, and at the same time a lesson of humility taught to each individual, when we are made to see that the universal necessary Laws, and pure IDEAS of Reason, were given us, not for the purpose of flattering our Pride and enabling us to become national legislators; but that by an energy of continued self-conquest, we might establish a free and yet absolute government in our own spirits."[86]

The third volume of *The Friend* contains, besides six miscellaneous essays, eleven essays "On the Grounds of Morals and Religion, and the Discipline of the Mind Requisite for a True Understanding of the Same." Eight of these eleven are broadly concerned with method, and the discussion "is confined to Method as employed in the formation of the understanding, and in the constructions of science and literature."[87] The constructions of science are most in evidence, and the symmetry between laws of reason and ideas of nature is made explicit. Coleridge's argument takes him beyond the understanding to reason in its secondary sense. The relations between primary and secondary reason render this strategy valid and intelligible.

Coleridge had good cause to advance gradually in presenting his discourses. An age that he saw as characterized by its adoption of mechanical philosophy needed aid in the ascent from sense to understanding, from understanding to secondary reason, and from secondary to primary reason. *The Friend* carried the reader up the first two steps, and indicated the existence of the third. The ascent of the third step was reserved for a subsequent publication, *Aids to Reflection*, which appeared in 1825. The aims of that work included the proposal and substantiation of the distinction between the understanding and reason – especially primary reason. Coleridge saw his own times as dominated by the understanding: "In no age since the first dawning of science and philosophy in this island have the truths, interests, and studies which especially belong to the reason . . . sunk into such utter neglect, not to say contempt, as during the last century."[88] He sought to show that to elevate the understanding above the reason was to lose the latter and to spoil the former.

The stress on the primary reason takes *Aids to Reflection* beyond the limits of this book. But even there, in a work on spiritual concerns, Coleridge constantly used scientific illustrations, discussing laws, causation, teleology, the ascent of life, instinct in bees and ants, and dynamic science. *Aids to Reflection* was published while he still maintained an active interest in the sciences, and two years before he directed the thoughts that he confided to his notebooks and friends almost exclusively into the realms of primary reasons. It sheds light on many of his explorations in science. I shall appeal to it often in the ensuing chapters.

3

TWO VISIONS OF THE WORLD

COLERIDGE, NATURAL PHILOSOPHY, AND THE PHILOSOPHY OF NATURE

A world of little things

Coleridge's objections to mechanism were all interdependent, but the imputation of atheism came to be the fundamental objection, just as theological preoccupations filled his later notebooks. At every state, he sought to keep his interpretation of nature, his philosophy, and his theology in step with one another. The tripartite organization of his first educational schemes was maintained over the years. "True Philosophy," he wrote in 1817, ". . . takes it's root in Science in order to blossom into Religion."[1] Science in itself was morally neutral, but it provided materials for the construction of philosophy, which in turn supported and was transcended by theology.[2] Atheism, in contrast, meant for Coleridge the end of reason and purpose in nature, the end of any possibility of rational knowledge or purposeful action. In an early note, Coleridge proposed to introduce into a poem "a dissection of Atheism – particularly the Godwinian System of Pride Proud of what? An outcast of blind Nature ruled by a fatal Necessity."[3] He believed that blind Nature and fatal Necessity followed inevitably from atheism and materialism, and he indentified two kinds of materialists and atheists. The first kind assumed that everything was material, that the atoms of matter were endowed with the passive attributes of figure and movability, and that thought and sensation arose from the accidental organization of atoms. The second kind added the active properties of self-movement and life to atoms and derived thought not from the mere arrangement of atoms, but from the aggregation of the properties essential to atoms. Neither kind had any need of God.[4]

Coleridge was confronting the results of a century's interpretation of Newton's writings, especially his *Opticks* and *Principia.* There is still controversy over Newton's views about the relation of force and power to matter and to God, contentious topics of eighteenth-century debate.[5] There were certainly grounds for debate. In his third rule of philosophizing in the *Principia*, Newton provided a rule for extrapolating from the knowledge of bodies derived from experience to assumptions about all bodies in nature, thus contributing to the possibility of a truly universal natural science. The essential properties of bodies that Newton had experimentally found to be incapable of intension and remission (scholastic terms for augmentation and diminution) were extension, hardness, impenetrability, mobility, and force of inertia. Because there was an analogy in nature, which is "accustomed to be simple and always consonant to herself," Newton concluded that the least parts of bodies were also extended, hard, impenetrable, mobile, and endowed with their own force of inertia. He gave, in short, what could be interpreted as an account of atoms, and underlined this account by describing it as the foundation of all philosophy.[6] It was possible to consider hardness, impenetrability, mobility, and inertia as the essential qualities of atoms, and to regard them as passive qualities. Newton's explicit exclusion of gravitation from this list of essential qualities, coupled with his demonstration of the universality of gravitation, appeared to underline the passivity of matter. On this view, activity in the world was maintained by God. Newton's "General Scholium" in the *Principia* strengthened this view by stressing God's dominion and activity. But Coleridge saw the assumption of God's absence from matter that was itself lifeless as opening the way to his first kind of atheist.

His second kind of supposed atheist either ignored Newton's separation of gravitation from the essential and passive properties of matter or attributed impenetrability and even extension to active principles in matter. In either case, the least parts of matter appear in this interpretation as active. A whole tradition of British Newtonians, from John Keill to Joseph Priestley,[7] had developed this interpretation, strengthened in their view by the thirty-first query to the *Opticks*, with its atoms endowed with a variety of forces. Now neither Keill nor Priestley would have endured being called an atheist. The latter, in his *Disquisitions on Matter and Spirit*, had approximated matter and spirit, narrowing the gap between the things of this world and divine activity, and making the cosmos more truly a continuum. Priestley was enormously satisfied by the idea of his most intimate connection with the deity, which seemed to flow from his theory. Coleridge, however, was simply shocked at a scheme that had no advantages over pantheism or Spinozism except inconsistency and that, in his judgment, led inevitably and directly to atheism.[8] The materialist who attributed activity to matter was a "Panhylist" who, "beat out . . . of mere matter yet unwilling to abandon it adopts the theory of Hylozoism . . . Life and Matter being the ground of all, all else are explained."[9] In Bristol, Coleridge read Ralph Cudworth, who had identified "the hylozoick Corporealist" as an atheist. Coleridge was prompted to jot down a note: "In the Essay on Berkley to speak of Sir Isaac Newton & other material theists . . . " He concluded his discussion of atheists by remarking of this second species that "almost all Surgeons, Chemists, & ~~Physician~~ ⟨Scotch⟩ Physicians" belonged to it. There were distinctions, grades of error, in atheism, and "Epicurean Atheism, or the godless lifeless Phantasm of crass Materialism," appeared as the worst.[10]

As early as 1796, in "The Destiny of Nations," Coleridge had condemned mechanistic theories invoking atoms and imponderable fluids. Atoms were unacceptable to him as anything more than a fiction, as useful for computation in chemistry "as xyz in Algebra." "But if they are asserted as real and existent, the Suffiction (for it would be too complimentary to call it a Supposition) is such and so fruitful an absurdity that I can only compare it to a Surinam Toad crawling on with a *wartery* of Toadlets on its back, at every fresh step a fresh Tadpole. The contradictions which it involves, were exposed by Parmenides, 460 years A.C., so fully as to leave nothing to be added." And, as he pointed out in his philosophical lectures in 1818, Epicurean atomism had no explanatory power, because it assumed that the world as it is at present constituted contains the causes of its present constitution.[11]

Coleridge considered that much modern chemistry had been vitiated by the improper use of John Dalton's chemical atomism. There were also extensions of atomism to account for forces by postulating subtle and imponderable fluids, presumably made up of smaller and finer atoms. The phenomena of electricity and magnetism were explained as resulting from the action of electrical and magnetic fluids, as in Franklin's theory of electricity. Thermal phenomena were attributed to the activity of a matter of heat, named *calorique* ("caloric") by Lavoisier.[12] All one needed to account for a distinct kind of activity in nature was to postulate a corresponding imponderable fluid as its cause. There were objections to this. Beddoes had advocated a chemistry of forces rather than of fluids, and Davy, having formerly advocated a chemistry of fluids, rebelled against it as contrary to reason.[13] Coleridge was increasingly intemperate against these fluids: "For an atom is a Pig with a buttered Tail, the instant you catch it, you lose it/ even the Caloric is in Jeopardy."[14] He believed that it was a grievous error always to hypostasize into fluids the active and productive principles in nature that he called powers; one could not produce power from mass that had no power but inertia. Decidedly, Coleridge concluded, only the despotism of fashion could persuade otherwise perspicacious men to designate powers as things. The root of the error lay in the attempt to "objectivize Powers by substituting the sensuous products as their representatives."[15] This attempt sought to arrive at a knowledge of causes and of the inmost nature of things by framing explanatory fictions derived from observation of the external properties of bodies.

The absurdity was palpable for Coleridge, and an obvious enough consequence of Newtonian thought. John Locke's epistemology likewise asserted that human knowledge came from the outsides of bodies, as apprehended by the senses and conveyed to the tabula rasa of a passive mind. Coleridge considered this arrogant and literal superficiality. He rebelled against the "despotism and disturbing forces of the senses," especially the tyranny of the eye in usurping reason in science and in philosophy.[16] He saw Locke as having taken corpuscular philosophy from the confines of science and inflated its emphasis upon passive externals into a system of so-called philosophy and of materialist psychology. Even Hartley's account of the mechanism of association had to be rejected as an offshoot of this mistaken growth. As for Condillac's writings, which were generously acknowledged by Lavoisier as fundamental for his method in chemistry, and which derived knowledge systematically from the senses, they were "hypothetical psychology on the assumptions of the crudest materialism, stolen too without acknowledgment from our David Hartley's essay on Man."[17] Coleridge warned against mistaking the conditions of a thing for its causes and essence. He lamented the follies of materialists, who, pretending that there were no mysteries, made everything mysterious, and merely obscured the difficulties inherent in such terms as action, space, matter, and cause.[18]

With convictions like these already forming within him in 1801, it is no wonder that Coleridge wrote to Poole disclaiming any intention of joining Locke's followers, "the party of the Little-ists."[19] There was an alternative to systems of science and philosophy based on that passivity of the mind which Coleridge found in Locke and in Newton. That alternative took him some years to explore and make his own. But his reading of Platonic and NeoPlatonic authors, his religious convictions, his feelings, and his critical thought all encouraged him to read carefully the metaphysical books that he brought back from Göttingen. A book that he may have read, perhaps before leaving for

Germany or shortly after his return, A. F. M. Willich's *Elements of the Critical Philosophy* (London, 1798), encapsulated the hopeful alternative in its first pages: "Kant remarked, that Mathematics and Natural Philosophy had properly become sciences by the discovery, that reason a priori attributed certain principles to objects; and he inquired, whether we could not also succeed better in Metaphysics by taking it for granted, that objects must be accommodated to the constitution of our mind, than by the common supposition, that all our knowledge must be regulated, according to external objects."[20] Coleridge studied Kant, whose works he admired. But he went beyond him. Kant viewed ideas as regulative: Coleridge followed Plato in regarding them as constitutive. Kant constructed a set of antinomies or paradoxes to show the limits of reason: Coleridge essentially identified Kant's reason with understanding, and refused to restrain the higher faculty of reason within the limitations of the understanding. He even argued that "the truths of Reason appear to the Understanding in the shape of contradictory pairs of propositions."[21] It followed for him that the dead ends of pantheism and atheism, born of corpuscular philosophy, had arisen simply because concepts of the understanding, with their inherent limitations, had been applied to ideas of reason.

Coleridge's extension, sometimes tantamount to contradiction, of Kant was at once his own and part of the wider enterprise of post-Kantian idealism. His admiration for Kant remained as striking as his differences from him, and he maintained that in Kant "is contained all that can be *learnt*."[22] What above all could be learned was the importance of logic and the active role of the mind in framing human knowledge. The latter was part of romanticism, the former part of classical philosophy. Schelling in Germany was the first to frame a philosophy of nature congnizant of both of these, and was thus the principal philosopher of romantic science. Coleridge read and used his works and those of his followers in formulating an alternative to the corpuscular approach to science. It is time to turn to the romantic philosophy of nature.

The active mind: science and romanticism

Coleridge had criticized Newton for making the mind a lazy looker-on, had rejected the Newtonian cosmos as a fragmented world of little things, and had condemned the sterile lifeless universe of mechanism with its inert atoms. Mechanical science took the living, moving, organic unity of nature and killed it by dissection and abstraction. Nature appeared to be an incoherent, mindless, and irrational chance agglomerate of lumps of matter. This was not Coleridge's complaint merely. It lay at the heart of the romantic rejection of mechanical science and of that mechanization of the world picture which had seemed to the Encyclopedists the greatest achievement of modern science. But "mechanical" became the great polemical adjective of the romantic movement in its contemplation of nature. Jean Paul, typically, spoke of the "all-powerful, blind, lonesome machine" of the universe.[23]

In place of mechanism, organic life became the central metaphor of romantic science. J. W. Ritter, who became the supreme authority on science for German romantics, viewed nature as the "All-Animal" (*All-Thier*).[24] Such a unified view rendered irrelevant the traditional distinctions among different fields of natural knowledge. The eighteenth century had seen, for example, efforts to realize the autonomy of chemistry as a distinct discipline.[25] But for romantic scientists, there were no adjacent fields; there were merely "various manifestations of the romantic spirit."[26] Wilhelm Nasse announced in 1809 that there was only one natural science, which was the recognition of unity.[27] The implication was daunting, but gleefully recognized. J. G. Rademacher, a physician, stated it with typical arrogance: "It is impossible to succeed in knowing a part of a whole without knowing the whole, for the part is not only connected with the whole but depends on it in a constant exchange of cause and effect in such a way that it attains its real significance and essence of character only through this exchange."[28] Ritter's metaphor of the *All-Thier* was grounded in his recognition that everything was related to everything else. The program thus generated for science was impossibly vast. One cannot comprehend infinity. The hubris of romantic science was foredoomed. Yet this consequence was far from apparent to the proponents of such programs as Ritter's. Indeed, it seemed to some that Ritter had achieved or was about to achieve the success of omniscience in his enterprise. Clemens Brentano, for example, urged his sister: "Write to Ritter as you would write to the universe. He is about to spell out the creation."[29] How could this ever be possible? There was only one hope of success, only one condition under which man could attain knowledge of nature within the framework of romantic thought, and this condition, as Ritter recognized, was that nature and man be alike.[30] Knowledge of nature and man's self-knowledge were thus interdependent.

Coleridge's interpretation of Platonism showed how classical philosophy could provide a foundation for this belief.[31] The acquirement of self-knowledge was a constant preoccupation for Coleridge, and also "everyman's interest and duty . . . : or to what end was man alone, of all animals, endued by the Creator with the faculty of self-consciousness?"[32] But Coleridge's world was a cosmos, not a chaos, a unity comprising man and nature. Self-knowledge required complementary knowledge of nature:

> 'Tis the sublime of man,
> Out noontide Majesty, to know ourselves
> Parts and proportions of one wondrous whole![33]

Romantic feeling pointed to a new mode of apprehending nature, to an active harmony between mind and nature. Such harmony was essential to the new understanding. In *The Soul in Nature*, the Danish physicist H. C. Oersted gave succinct expression to this view: "The world and the human mind were created according to the same laws. If the laws of our reason did not exist in Nature, we should try in vain to force them upon her; if the laws of nature did not exist in our reason, we should not be able to comprehend them."[34] In England, Thomas Taylor's translations of Plato, Proclus, and Plotinus reinforced English romantic authors in viewing the mind as active in nature, and influenced their choice of imagery.[35] It is worth remarking that some of Taylor's translations appeared while Coleridge was at Cambridge. Then, in Germany, came attempts to combine the romantic universe of feeling with the development of critical philosophy, to arrive at a new philosophy of nature. Kant's works were the starting place for the Germans, as they had been for Coleridge, in an irony that would scarcely have been appreciated by Kant himself, pietist, rationalist, Newtonian, and "antiromanticist *par excellence*."[36]

***Naturphilosophie* and the life of nature**

Coleridge was concerned to understand the role of mind and self-consciousness in arriving at a system of science and the philosophy of nature. Fichte and Schelling similarly began their investigations from Kant's doctrine of the self.[37] They interpreted this doctrine as revealing that experience arose from the interaction of the knowing mind, the subjective component, and the thing that is known, the objective component. They went on to argue that it followed from this doctrine that there were only two alternative systems of metaphysics, arrived at by abstracting from either the objective or the subjective component. If, said Fichte, the philosopher "abstracts from the former, he gets, as the explanatory ground of experience, an intelligence-in-itself, that is, intelligence abstracted from its relation to experience;

if he abstracts from the latter, he gets a thing-in-itself, that is, in abstraction from the fact that it is presented in experience. The first procedure is called idealism; the second dogmatism."[38] Dogmatism thus took Kant's noumena as fundamental. Fichte rejected this approach because it seemed to him to make the category of cause applicable to the nonempirical self, which could not then be free. The alternative approach rejected things-in-themselves and, by taking mind or intelligence as its foundation, transformed Kant's philosophy into a thoroughgoing self-consistent idealism. If there were no noumena, what was represented in our perceptions? Representations, said Fichte, were "productions of the intelligence, which must be presupposed to explain them." The self, the "I" in "I think," was the source of being and knowledge. It posited its own being, and also the world outside it, which was its own creation.[39]

Coleridge understood this in reading Fichte's *Wissenschaftslehre* of 1794, for all the ridicule he gave it in writing to Dorothy Wordsworth in 1801. Fichte's "I" was constitutive through its activity. In the *Biographia Literaria* this doctrine came in for tribute and criticism: "FICHTE's *Wissenschaftslehre* . . . was to add the key-stone of the arch [of Kant's philosophy]: and by commencing with an *act*, instead of a *thing* or *substance*, Fichte . . . supplied the idea of a system truly metaphysical . . . But this fundamental idea he overbuilt with a heavy mass of mere *notions*, and psychological acts of arbitrary reflection. Thus his theory degenerated into a crude egoismus, a boastful and hyperstoic hostility to NATURE, as lifeless, godless and altogether unholy."[40] Fichte had made a start in replacing the "thingical" philosophy with a dynamic philosophy, having substituted act for thing.[41] But he had still not provided a philosophy that accorded with the romantic vision of the world as a living organic unity. Schelling, who began as a disciple of Fichte and like him followed Kant while subverting his doctrines, attempted to provide an idealistic philosophy that would accord with this vision. Schelling's dynamic philosophy was in major respects congruent with the romantic view of nature; this congruence helped to inspire Coleridge to sketch an outline of the history of science contrasting dynamic science with mechanical science.[42]

Schelling followed Fichte in rejecting things-in-themselves. He also recognized the existence of subjective or ideal and objective or real components in knowledge, but he insisted that there was ultimately no division between them.[43] His attempt at resolving the apparent duality came through his formulations of transcendental idealism and of *Naturphilosophie*, and through his argument that these were complementary. In formulating his *System des Transcendentalen Idealis-*

mus (1800), Schelling started with self-consciousness as the foundation of philosophy. He then proceeded to deduce nature from absolute self-consciousness. Nature was thereby comprehended in absolute mind – indeed, generated by its activity. Here was a striking opposition to what Coleridge saw as the fundamental flaw in Newton's system, the passivity of mind. Here too was an advance beyond Fichte, for whom nature had been only the not-I.[44] And even beyond this, Schelling argued that mind produced nature through productive imagination – a doctrine powerfully attractive to Coleridge and much used by him.

Naturphilosophie, the second part of Schelling's philosophy, attempted to explain the ideal by the real. Schelling asserted that nature's products necessarily show design. "But if nature can produce only the regular, and produce it from necessity, it follows that the origin of such regular and design-evincing products must again be capable of being proved necessary in nature, regarded as self-existent and real."[45] *Naturphilosophie* in this sense is realistic; it seeks to explain the physical world "by reference to concepts and principles discoverable in, and appropriate to, the physical world, as opposed to concepts and principles which are discoverable in, and appropriate to, the human mind."[46] *Naturphilosophie* is not experimental physics, but speculative physics, whose aim is to discover the structure of nature through the modes of its activity, the forms of the laws by which it is governed.

Naturphilosophie and transcendental idealism are brought together through Schelling's doctrine of identity (*Identitätslehre*).[47] In knowledge, the subjective and objective components are ultimately identical; self-consciousness and knowledge are possible only because of the absolute identity of subject and object. In the world there is likewise an absolute identity of mind and nature. The different emphases in transcendental idealism and *Naturphilosophie* arise because philosophical inquiry can be pursued in different ways that, in consequence of identity, turn out to be complementary investigations of different aspects of an ultimate cosmic unity. The doctrine of identity is Schelling's solution to the problem of the ideal and the real; he considered that it raised philosophical thought to a new level. It could also be seen as a variant of Spinoza's solution. McFarland has demonstrated that Schelling took Kant's thought, which was opposed to Spinoza, and turned it into an anti-Kantian variant of Spinozism. *Naturphilosophie* was the *Spinozismus* of physics.[48]

Schelling considered nature to be productive activity, and physical objects to be products. Spinoza's *natura naturans* and *natura naturata* are precise analogues, and just as these are one for Spinoza, so nature as productivity and product are one for Schelling. Theory deals with the former, empiricism with the latter. The transition from becoming to being, the evolution of productivity into product, is continuous. Potentially infinite productivity must be limited, in order to generate finite products. Thus productivity must have within itself an opposing restraint, and nature as productivity, *natura naturans*, has within itself an essential polarity.[49] The task for *Naturphilosophie*, said Schelling, the "first principle of a philosophical system, is to *go in search of polarity and dualism throughout all nature*."[50]

Schelling educes further general characteristics of nature from this argument. First, because nature is one with mind, it is rational in its productivity. *Naturphilosophie* seeks to show the genetic character of nature through a correspondingly rational deduction of a sequence of productive laws. Secondly, nature as productivity, *natura naturans*, is organic in its activity. The unity of mind and the identity of mind and nature render the organism of nature a unified one. Schelling's *Naturphilosophie* accords well with his thesis in *Von der Weltseele* (1798): "As soon as our investigation has risen to the idea of nature as a whole, the opposition between mechanism and organism, which has so long hindered the progress of the natural sciences, will disappear." In this conception of nature, "a general organism is the prerequisite for everything mechanical."[51] "Is not mechanism . . . the negative of organism? . . . Must not organism be prior to mechanism, the positive prior to the negative? . . . So our philosophy cannot proceed from mechanism (as the negative), but must proceed from organism (as the positive)."[52] Because nature is organic, productive, and unified, the sciences should be genetic and synthetic. They should also incorporate in the forms of their laws and fundamental concepts the mode of nature's productivity. They should reflect the dynamic aspect of polarity. Everything in nature is constantly striving forward, because of a positive principle, the first force of nature; "but an invisible power brings all appearances [*Erscheinungen*] in the world back in the old circular course," because of a negative principle, the second power of nature.[53] Here, in the guise of opposite forces, was Schelling's fundamental polarity in nature. There is an analogy here with the forces of attraction and repulsion that for Kant constituted matter, and Schelling acknowledged this.[54] The dynamism of polar opposites in constant strife was central and constitutive in Schelling's philosophy; it characterized *Naturphilosophie*. The result was a view of nature as a continuous flux of opposed forces in dynamic tension, whose very opposition and ultimate identity or dynamic synthesis constituted matter with its powers; qualities of matter arose from powers. All nature was ultimately one, as natural philosophy could show. The scientist was concerned to demonstrate the interrelations of polar forces, the dynamism of every department of nature, the underlying unity and continuity of nature.

Schelling developed *Naturphilosophie* so as to deduce the dynamic process of nature and construct the forms of matter. He published his deductions in his journal for speculative physics, which Coleridge read and annotated, along with other works by Schelling.[55] Coleridge was attracted to Schelling's ideas because they seemed at first to reconcile Kant with Spinoza, to bring together "I am" with "it is." Coleridge came to recognize that Schelling's doctrines were pantheistic, and he struggled to free himself from them in his inability fully to reject or to accept pantheism, to which he was intellectually opposed and emotionally attracted.[56] He developed a philosophy differing significantly from Schelling's. He refused to make nature absolute, but he also acknowledged that "in Schelling's 'NATUR-PHILOSOPHIE' and the 'SYSTEM DES TRANSCENDENTALEN IDEALISMUS,' I first found a genial coincidence with much that I had toiled out for myself, and a powerful assistance in what I had yet to do."[57] Schelling had furnished him with a philosophy that, however flawed, put mind and life back into nature, and showed the unity of mind and nature including man with his creative imagination. Schelling had formulated a philosophy tending in the same direction as Coleridge's thought and feeling, and in so doing had shown him the way to a dynamic science that would oppose and topple sterile, dead, mechanical science.

Coleridge may have been the first man in England to see this.[58] But he was not a scientist, and he waited until scientists following Schelling had tried to build a dynamic science, a science based upon concepts of productivity and polarity. Some scientists, notably in Germany, sought to apply Schelling's deductions not only to a reinterpretation of the known facts of science, but to the prediction of new facts and even to the pursuit of new research programs directed to the extension of the correlation between speculative physics or *Naturphilosophie* and experimental and empirical physics. Ritter pursued such a program. He observed the polar phenomena of galvanism and set about demonstrating galvanic properties throughout nature, and galvanic sensitivity from minerals to man.[59] Ritter's enthusiasm, especially in his later years, made him increasingly careless of the distinction between speculative and empirical physics. Coleridge, for one, was critical of this failing.[60]

English scientists had little time for Ritter and his speculations; their science was essentially empirical, whether directly in Locke's train or acknowledging Scottish commonsense philosophy. The French were in general quite as hostile to *Naturphilosophie* as the English were, and they were skeptical of discoveries inspired by German philosophy.[61] But if they lacked German method, they had their own system and in the life sciences were the great classifiers in early nineteenth-century Europe. The sciences might be international, but styles of science were at that time markedly national.

Science, history, and principles

Coleridge made much of the contrast between dynamic science, largely stemming from Germany, and what he called the mechanical science of England and France. He used it as a form of shorthand to illustrate the opposition between philosophical traditions as well as that between contemporary philosophical positions, and even to diagnose the state of society and the spirit of the age in different nations. In 1817 he wrote to Lord Liverpool, to that nobleman's evident puzzlement, that the acceptance of chemical atomism in England and of mechanical atomism in France "determine the intellectual character of the age with the force of an experimentum crucis." Coleridge was fully aware that the world at large was scarcely concerned with the philosophy of nature or cognizant of its implications, but he argued that his reading of history showed "that the Taste and Character, the whole tone of Manners and Feeling, and above all the Religious (at least the Theological) and the Political tendencies of the public mind, have ever borne such a close correspondence, so distinct and evident an Analogy to the predominant system of speculative Philosophy, whatever it may chance to be, as must remain inexplicable, unless we admit not only a reaction and interdependence on both sides, but a powerful, tho' most often indirect influence of the latter on all the former."[62]

The need for a reform of society in a manner avoiding the excesses of revolution at home and the oppression of imperalism abroad moved Coleridge to address this homily to the prime minister. Such reform could come only if the higher classes of society, the gentry and clergy, discerned and acted upon true principles. Now principles, guides to thought and action, derived from philosophy. "As long as the principles of our Gentry and Clergy are grounded in a false Philosophy, . . . all the Sunday and National schools in the world will not preclude Schism in the lower & middle Classes. The predominant Philosophy is the key note."[63]

Coleridge's letter to Lord Liverpool came just one year after his publication of *The Statesman's Manual; or, The Bible the Best Guide to Political Skill and Foresight: A Lay Sermon Addressed to the Higher Classes of Society*. In that work Coleridge had advocated a philosophical reading of history for an understanding of the rules of action governing society. He argued that the mainspring of social change drew its energy from the theories of intellectuals who often seemed isolated and remote from practical affairs. Speculative philosophy was, he admitted, the province of the few. "Yet it is not the less true, that all the *epoch-forming* Revolutions of the Christian world, the revolutions of religion and with them the civil, social, and domestic habits of the nation concerned, have coincided with the rise and fall of metaphysical systems. So few are the minds that really govern the machine of society, and so incomparably more numerous and more important are the indirect consequences of things than their foreseen and direct effects."[64] He reinforced this argument by claiming that in times of rapid change and social unrest, abstract notions came very close to the feelings and motives of the populace: "At the commencement of the French revolution, in the remotest villages every tongue was employed in echoing and enforcing the almost geometrical abstractions of the physiocratic politicians and economists."[65] The evils of the revolution and the Terror could all be traced to the espousal of false philosophy, and to the neglect of true philosophy and of the lessons of history. Coleridge in this mood viewed history simply as an exercise in identifying principles and the results of adopting or neglecting them. Bacon, James Harrington, and Machiavelli were among the authors whom he recommended for the edification of the higher classes. Alas, wrote Coleridge, they were seldom read in Regency England, perhaps for the very reason that Hume, Condillac, and Voltaire with their different principles were read.[66]

Principles, rules of conduct, had the character of first causes – for example, in morality. They were powerful precisely because, like the first cause in nature, they initiated an ever-extending series of observable secondary causes and effects. "Every principle is actualized by an idea; and every idea is living, productive, partaketh of infinity, and (as Bacon has sublimely observed) containeth an endless power of semination. Hence it is, that science, which consists wholly in ideas and principles, is power."[67] Coleridge pointed to Edmund Burke as a statesman who acted according to conscious principles and who was therefore a scientific statesman and "a seer. For every *principle* contains in itself the germs of a prophecy; and as the prophetic power is the essential privilege of science, so the fulfilment of its oracles supplies the outward . . . test of its claim to the title."[68] Science was accordingly particularly valuable for diagnosing the principles of an age and of a nation, and the value of the history of science lay in extending the diagnosis beyond the confines of the present age. This extension, stemming in general from "the perusal of our elder writers," was necessary. "It will secure you from the narrow idolatry of the present times and fashions, and create the noblest kind of imaginative power in your soul, that of living in past ages; – wholly devoid of which power, a man can neither anticipate the future, nor ever live a truly human life, a life of reason in the present."[69]

Referring men's opinions to principles was the foremost and avowed aim of *The Friend*, which appeared in a revised three-volume edition in 1818. In spite of his lack of liking for the bones of history, Coleridge considered that he had read "most of the Historical Writers" as early as 1796, and was besides "*deep* in all the out of the way books, whether of the monkish times, or of the puritanical aera." He used his historical reading to identify principles and to illustrate their consequences, and constructed his own version of the history of ideas and the history of science to reinforce his conclusions. The contrast between mechanism and dynamism informed his accounts, and was correlated in them with the history of philosophy. Early philosophers – those who wrote before the scientific revolution – had cultivated metaphysics to the neglect of empirical psychology, whereas later philosophers had reversed this emphasis. Coleridge regarded both approaches as equally unbalanced and removed from "true philosophy." These motives and generalizations inform Coleridge's version of the history of science as a part of cultural history symbolizing the whole.[70]

Coleridge's history of science: dynamism versus Anglo-Gallic mechanism

Coleridge began his history of science in *The Friend* of 1818 with the book of Genesis, which he viewed as directed to the cultivation of intellectual and spiritual faculties. He contrasted this with the sensuality of those who sought to derive knowledge from the appearance of things: "They built cities, invented musical instruments, were artificers in brass and in iron, and refined on the means of sensual gratification . . . they became idolaters of the Heavens and the material elements." Then came ancient Greece, representing "the youth and approaching manhood of the human intellect," taking up into its mysteries some of the poetry of the Hebrew scriptures, and effecting a "restoration of Philosophy, Science, and the ingenuous Arts." But the almost miraculous intellectual greatness of the Greeks was not matched by a corresponding application of the mind to the practical investigation of nature. Greek excellence was of the mind and its self-reflection; where mind sufficed, in art and science, the result was "an almost ideal perfection." But in confronting the sensible world in its chemical, mechanical, and organic forms, the Greeks were strikingly unsuccessful. The contrary imbalance was demonstrated in the third period, that of the Romans, who rendered material the ideas of the Greeks, and displayed a purely practical talent.[71]

Coleridge has little to say of the history of science from the Greeks to the Renaissance, where he resumes his account with praise for the alchemists, whose enterprise he held to be the true goal of chemistry: "There must be a common law, upon which all can become each and each all."[72] He read some of the medico-chemico-surgical works of Paracelsus.[73] But Kepler was the sixteenth-century scientist whom he most admired. Kepler was, indeed, the object of widespread admiration in the early nineteenth century, especially in Germany; and both S. Vince's *Complete System of Astronomy* and Robert Small's *Account of the Astronomical Discoveries of Kepler*[74] showed that this feeling was not confined to Germany. Coleridge contended that Kepler, whose ideas Newton merely completed in framing his system of gravitation, was the founder of astronomy as a science.[75]

> Galileo was a great genius, and so was Newton, but it would take two or three Galileos and Newtons to make one Kepler. It is in the order of Providence, that the inventive, generative, constitutive mind – the Kepler – should come first; and then that the patient and collective mind – the Newton – should follow, and elaborate the pregnant queries and illumining guesses of the former. The laws of the planetary system are, in fact, due to Kepler. There is not a more glorious achievement of scientific genius upon record, than Kepler's guesses, prophecies, and ultimate apprehension of the law of the mean distance of the planets as connected with the periods of their revolutions round the sun.[76]

Coleridge stressed the contrast between Kepler and his successor Newton. He also contrasted Kepler to Francis Bacon, turning from one to the other as from sunshine to gloom. Whereas Kepler's imaginative genius won Coleridge's affection, Bacon earned merely admiration. His was a complex mind that has been so variously portrayed as to seem more than Janus-like. The early members of the Royal Society all but canonized him, and nineteenth-century popularists, mindful more of Bacon's experiments of fruit than of his experiments of light, portrayed him as the founding utilitarian empiricist.

This is the Bacon of the *New Atlantis*, whose sages are akin to Coleridge's clerisy. But Bacon was also the author of works that, like his *Sylva Sylvarum*, were less useful to the historiographer of the scientific revolution. Coleridge was more sympathetic to such works than were most of his contemporaries, and insisted that one should not, for example, "bring the belief in the physical powers of the Will & Imagination as proofs of disqualifying Credulity in Bacon."[77] If one took care in reading Bacon historically, seeking to understand each word as he intended it and recognizing the rhetorical structure of his arguments, then many apparent crudities and contradictions disappeared. Here Coleridge displayed in historiography the delicacy he too often confined to literary criticism. He then went on to propose a further degree of clarification through concentrating on Bacon's principles separated from their often erroneous applications – and went on to complete this now strikingly unhistorical procedure by dismissing those passages in which Bacon was not true to himself. This mixture of sensitivity and coercion yielded a most impressive Bacon, "the Founder of a revolution, scarcely less important for the scientific, and even for the commercial world, than that of Luther for the world of religion and politics."[78]

Bacon's importance for Coleridge lay predominantly in his statements about scientific method. Coleridge interpreted Bacon's concentration on "the material pole" as containing within itself a counterbalancing intellectual opposite. For all Coleridge's forcing Bacon into his own image, there is some truth in this view. But this truth was increasingly neglected by the natural philosophers associated with the new Royal Society of London – and how Coleridge grudged such men and their heirs the title of "philosopher." As early as 1804, Coleridge had written complaining of Davy's misuse of that honored word: "I have met with several genuine Philologists, Philonoists, Physiophilists, keen hunters after knowledge and Science; but Truth and Wisdom are higher names than these – and *revering* Davy, I am half angry with him for doing that which would make one laugh in another man – I mean, for prostituting and profaning the name of Philosopher, great Philosopher, eminent Philosopher &c &c &c to every Fellow, who has made a lucky experiment."[79] William Whewell tells us how Coleridge, in the year before his death, stood up in Cambridge to forbid the British Association for the Advancement of Science the use of "philosopher" to describe any student of the material world. The word "scientist" was first proposed on that occasion, though it was not then found generally palatable.[80]

The scientists of the Royal Society generally overlooked the ideal pole in stressing the material. They followed Bacon's advice to collect and tabulate particulars as the materials of a natural history, advice that Coleridge judged to be quite independent of Bacon's "inestimable principles of scientific method." Robert Hooke, curator of experiments to the Royal Society, furnished Coleridge with a fair instance of the Baconian natural historian confronting separate things and single problems. Coleridge was appalled at the catalogue of miscellaneous information that Hooke in his *Posthumous Works* proposed as the necessary preliminary knowledge that would prepare the naturalist to form theories. The history of Hooke's "multifold inventions, and indeed of his whole philosophical life," Coleridge commented, "is the best answer to the scheme – if a scheme so palpably impracticable needs any answer."[81] And he quoted from Hooke's list, which ran the gamut of every artisan's trade and embraced the activities of dancing masters, seamsters, butchers, barbers, and makers of marbles.

Coleridge viewed Hooke's approach to nature as derived from a caricature of Bacon when least true to himself and devoid of informing principle.[82] It dealt with the externals of nature, took them one by one in isolation, judged of them through the senses, and led to an image of nature as a world of little things. Coleridge had called Locke the founder of the "Little-ists"; but Locke was making explicit a philosophy that came to be seen as implicit in the new science of the seventeenth century. Coleridge's abuse of Hooke is part of his attack upon mechanism in all its guises: superficiality, submission to the senses, materialism, purposelessness – even, ultimately, atheism.

There was, however, one mechanical philosopher toward whom Coleridge was ambivalent: Robert Boyle, who had done more than any other man in England to render corpuscular philosophy respectable and to purge it of atheism. Boyle was a friend of Locke's; he was also the founder of the Boyle Lectures, devoted to expounding the relations between science and religion. In a marginal note in a volume of Jean Baptiste de Boyer's cabbalistic correspondence, Coleridge criticized Spinozism for considering God to be nature, remarked on the wisdom of using nature as a general term, and added that Boyle had shown that it was dangerous to regard nature as being real and the product of an unintelligent power.[83] Such perceptions earned Boyle his place with Shakespeare and Milton among "the great living-dead men of our isle."[84]

In Coleridge's scenario for the rise of modern science, Kepler had discovered laws of planetary motion and had had some ideas about the dynamics of the solar system.[85] Coleridge chose to interpret these as a full and prior conception of gravitation. Bacon had striking views on scientific method. When he was true to himself, he seemed to Coleridge to be the British complement to Plato; at other times, which Coleridge judged to be foreign to Bacon's true genius, Bacon propounded views that Coleridge caricatured in his account of Hooke, and that Swift ridiculed in his account of the Laputan academy. And shortly afterward Thomas Hariot and Boyle in England, and Pierre Gassendi and Descartes in France, with "the restoration of ancient geometry, aided by the modern invention of algebra, placed the science of mechanism on the philosophic throne." Then came Newton, whose "sublime discoveries . . . and . . . his not less fruitful than wonderful application, of the higher mathesis to the movements of the celestial bodies, . . . gave almost a religious sanction to the corpuscular and mechanical theory."[86]

This apparent praise of Newton was unusual from Coleridge, and was indeed not spontaneously his own, but was taken from Steffens.[87] Far more typical was Coleridge's reaction to Pope's lines,

> Nature and Nature's laws lay hid in night:
> God said, *Let Newton be!* and all was light.

"I have been *un-English* enough," Colerdige complained, "to find in Pope's tomb-epigram on Sir Isaac Newton nothing better than a gross and wrongful falsehood conveyed in an enormous and irreverent hyperbole."[88] Just what was Newton's place in Coleridge's view of the history of science? He was the successor if not the heir to Kepler, Bacon, and Boyle. He expanded and organized Kepler's genius, merely developing Kepler's idea of the solar system.

We have already encountered Coleridge's estimate of the relative merits of Kepler and Newton; he went on forcibly to underline his estimate: "We praise Newton's clearness and steadiness. He *was* clear and steady, no doubt, whilst working out, by the help of an admirable geometry, the idea brought forth by another." Newton, then, took Kepler's form of the solar system, and transformed it into a mathematical structure lacking the life and purpose of Kepler's vision. Kepler's universe was one and alive.[89]

Coleridge saw Newton's debt to Bacon as less gross.[90] The colors that Newton identified as individually pure but severally constituting white light are akin to Bacon's simple natures. The third rule of philosophizing and the *Opticks* were together responsible for most of the consequences to which Coleridge most objected. He had annotated the 1721 edition of the *Opticks*, dismissed its first book as unsatisfactory, and rejected the theory therein as superficial and false.[91] Its fragmentation of color was part and parcel of the reduction of nature to a world of little things. Coleridge's related general objections to the mechanical consequences of Newton's method in the *Principia* have already been sufficiently indicated. Newton's philosophy was not the simple mechanism that Coleridge attacked, but Newton's ether did seem conformable to Boyle's corpuscular theory, just as Locke's doctrine of primary and secondary qualities could be seen as an extension of Boyle's explanation of qualities by the bulk, motion, form, and texture of corpuscles. Newton, said Coleridge, had his ether, and could not conceive the idea of a law embodying necessary cause.[92] Coleridge believed that Newton had tried to explain nature mechanically, that is, to explain mechanism by itself. But the only unity and totality that could contain within itself "the causative principle of its comprehended distinctions."[93] would be equal to God. Newtonianism would thus lead to atheism or pantheism.

Coleridge believed that Newton, who by the success of his works had excelled Descartes in enthroning mechanism as supreme in science, had by that same success burdened his followers with an intolerable dilemma arising from the very nature of mechanism. Nevertheless, mechanism "became synonymous with philosophy itself. It was the sole portal at which truth was permitted to enter. The human body was treated of as an hydraulic machine, the operations of medicine were solved and alas! even directed by reference partly to gravitation and the laws of motion, and partly by chemistry, which itself, however, as far as its theory was concerned, was but a branch of mechanics."[94] Science, and indeed all of intellectual life, followed the fashion – for perhaps the first time science was thus debased by fashion.[95] Mechanical philosophy was a philosophy of the senses, taking its tone more from Sterne's sentimental philosophy than from reason. The age of mechanism was "the Epoch of the Understanding and the Senses," and it produced among its adherents an evasion of reality and morality.[96]

Coleridge on national styles of science: France, England, and Germany

Coleridge regarded France as the country in which intellectual and social life had been most fully subjected to the forms of mechanical philosophy. Lavoisier's chemistry, with its plurality of elements and its imponderable fluid of caloric, was one consequence of this subjection.[97] The French Revolution and the Terror that succeeded it were for Coleridge no less the consequences of adherence to mechanism. He identified among the causes of France's upheavals "the predominance of a presumptuous and irreligious philosophy, . . . the extreme over-rating of the knowledge and power given by the improvements of the arts and sciences, especially those of astronomy, mechanics, and a wonder-working chemistry" – precisely those sciences most imbued with the spirit of mechanism.[98]

Coleridge was a patriotic Englishman who had lived through the years of the Terror and of the Napoleonic wars, and now detested France as "the most light, unthinking, sensual and profligate of the European nations." He came to see the version of mechanical philosophy manifested in Lavoisier's chemistry as peculiarly although not exclusively French, and called it "psilosophy" – "from the Greek, psilos slender, and Sophia Wisdom, in opposition to Philosophy, the Love of Wisdom and the Wisdom of Love." Alas, England was also given over to this slender wisdom; Lavoisier's caloric had its counterpart in Dalton's atomism. Coleridge conflated both into "Anglo-Gallican" chemistry, regretting that England was characterized not only by Shakespeare, Milton, and Bacon, but also by Locke, Pope, and Priestley. The "spiritual platonic old England" stood in opposition to "commercial G. Britain," and Coleridge's identification with the former was as firm as his alienation from the latter.[99]

He deplored the corruption of English thought by France, refusing to accord any virtues to the latter nation. He would admit no exceptions. He was a great admirer of Cuvier's work, and the "Essays on the principles of method" in *The Friend* of 1818 seem indebted to some of Cuvier's statements about scientific method. Now Cuvier was the doyen of the life sciences in France, and a stern critic of German metaphysical sciences, which he condemned for contributing nothing to the explanation of positive facts.[100] Yet his views on the relation of parts to wholes were similar to Coleridge's, and the discussion of classification in Cuvier's *Lectures on Comparative Anatomy* was echoed very closely by Coleridge in 1828. Coleridge was happily able to point out that Cuvier was not born in France, was not of "unmixed French extraction," and had studied "in a very different school of methodology and philosophy than Paris could have afforded." Cuvier's apparent hostility to German thought merely betrayed the suppression of his true feelings, consequent on his French readers' lack of philosophical sympathy. Coleridge was able to save his generalization about France.[101]

England's enslavement to "psilosophy" and estrangement from philosophy were nowhere more apparent than in her scientific societies. The Royal Society, the foremost scientific society of the nation, was also the prime offender in its subjection to mechanical philosophy. Davy's apparent apostasy from dynamism was accompanied by successive honors from the Royal Society, whose fellows elected him their president in 1820. Coleridge's disillusionment with Davy lent added point to his scorn for the society. In the summer of 1817 he had hopes that the London Philosophical Society might be an effective counterweight, "by pursuing the directly opposite course to that which the Royal Soc. has taken for the last 30 years." The plan clearly failed, because in December 1817, Coleridge, informing his close friend Joseph Henry Green of his forthcoming lecture on the principles of experimental philosophy, explained that it would take place "at the London Psilosophical Society in Flower de Luce Court, or thereabouts." Anglo-Gallic chemistry had subverted yet another institution.[102]

The Royal Institution was in no better state. Coleridge had lectured there in 1808, had failed to complete his course of lectures, had engendered frustration in the Royal Institution, and had felt frustrated by it. At that time he honored it for sponsoring Davy's electrochemical researches, while condemning it for injudicious management in all other things.[103] Davy's successor at the Royal Institution was William Thomas Brande, whose authoritative *Manual of Chemistry* (London, 1819) became one of Coleridge's principal sources in science. The *Manual* was comprehensive, well-organized, and firmly within the Anglo-Gallican tradition with its atoms, ether, and caloric. Coleridge was indignant at what he considered to be insularity masquerading as cosmopolitanism; it meant "good bye to the only sane, ⟨sound,⟩ sensible ~~and~~ solid *European* Philosophy, which, as suits an Age of *Amusement* and accords with the genius of its Birth place (*Abdera*, a *European* City in *Crim Tartary* or thereabouts) is *au desespoir* and good for nothing *sine Rebus in Rebus*." Abdera was the birth place of Democritus, one of the earliest Greek atomists, and was also notorious for the dullness of its inhabitants. "What," asked Coleridge in concluding his tirade, "What would the Lady Chem'stesses [*sic*], the fair auditory of the Royal Institution, do without *Things* in *Things*? Why, nothing could come of it. Much better, disguise Defeat in Devotion, and admit that these *things* 'remain among the mysteries of Nature, which, doubtless for the wisest purposes, are hidden from our view.' Sham, Flam, and Saintship, Humbug and holy-cant, rather than plead guilty of the horrid Heresy of Zoodynamism, and the *Life* of Nature."[104]

The life of nature was Coleridge's answer to the mechanists on the one hand, and on the other hand to Schelling's followers, who made nature absolute. He had arrived at his answer by 1819, after several abortive attempts vitiated by holding too faithfully to *Naturphilosophie*. But his position was closer to German than to British philosophy. As early as 1807 he had contrasted "Psilosophia Gallica" with "philosophia Teutonica."[105] In his letter to Lord Liverpool, and in an almost contemporary letter to the Swedenborgian C. A. Tulk, he referred to Germany as the only country where a man could exercise his reason without being supposed to be out of his senses. Germany was the only land where philosophy was still in repute, and he regarded the worst speculative German philosophy of the preceding half century as better than the best British empiricist philosophy produced in that period.[106] If science outside Germany were to recover its integrity, it could do so only through a revolution in philosophy in which dynamism would subvert mechanism.

Toward a new dynamism in science

The natural sciences in the early decades of the nineteenth century and even in the closing years of the eighteenth century seemed as if they would contribute significantly to such a revolution. By 1825, Coleridge was persuaded that mechanism had received "a mortal blow from the increasingly dynamic spirit of the physical sciences now highest in public estimation." And with his vision of the unity of intellectual life, Coleridge saw this dynamic spirit spreading from the sciences to every field of modern thought. Davy's work in electrochemistry was susceptible of a dynamic interpretation. Oersted's electrochemistry was avowedly dynamic, and had been followed by his discovery of the interaction of electrical and magnetic forces – this was a triumph of dynamism. John Hunter's theory of life had conferred outstanding unity on his physiological researches. Modern science was showing nature in all its complex unity and flux. The most accurate chemical analyses showed relations between the most disparate substances, and displayed at the same time the sheer inadequacy of chemistry to account for the life of nature. When Berzelius, a superb chemical analyst, listed the constituents of blood, or when Charles Hatchett, a meticulous chemist whom Coleridge admired, analyzed other animal substances, they revealed merely a momentary aspect of that flux. Nature in its totality was organic; its parts with their distinctive forms were "evolved from the invisible central power," and grew by assimilation rather than passive accretion:

> The germinal power of the plant transmutes the fixed air and the elementary base of water into grass or leaves; and on these the organific principle in the ox or the elephant exercises an alchemy still more stu-

> pendous. As the unseen agency weaves its magic eddies, the foliage becomes indifferently the bone and its marrow, the pulpy brain, or the solid ivory. That what you see is blood, is flesh, is itself the work, or shall I say, the translucence, of the invisible energy, which soon surrenders or abandons them to inferior powers, (for there is no pause nor chasm in the activities of nature) which repeat a similar metamorphosis according to their kind; these are not fancies, conjectures, or even hypotheses, but facts; to deny which is impossible, not to reflect on which is ignominious. And we need only reflect on them with a calm and silent spirit to learn the utter emptiness and unmeaningness of the vaunted Mechanico-corpuscular philosophy.[107]

Mechanism in Coleridge's schematic history of science had ousted the dynamism of Kepler and Bruno, had become the dominant philosophy in the second half of the seventeenth century, and had held unopposed sway for a full century.[108] Dynamism was first reasserted by Kant in his account of the *vis viva* controversy in mechanics, *Gedanken von den wahren Schätzung der lebendigen Kräfte* (1747).[109] Then the advent of dynamism in the natural sciences and of its philosophical complement in the work of Schelling prepared the way for the general restoration of dynamism that Coleridge so greatly desired. Germany was the nation with the greatest metaphysical genius, and metaphysical dynamics was thus not surprisingly a German creation. It furnished a foundation for physics,[110] making inroads in England almost exclusively through the superstructure of the sciences.

It could not be otherwise, because Coleridge saw England as barren ground for metaphysical seed. He identified Richard Saumarez's *New System of Physiology*, published in 1797, as the work marking the instauration of the dynamic philosophy in England. Saumarez had written the work to confute the Brunonian doctrine of excitability, offering in its place a system that took for its principle the power of life and stressed final causes throughout nature. A subsidiary meaning of *dynamic* emerges here. Coleridge generally used it to stress the role of powers and productivity; but there was also a related Kantian use. A. F. M. Willich, in his *Elements of the Critical Philosophy*, explained that Kant called "a *synthesis* dynamical, where the things combined necessarily belong to one another, but must not necessarily be of a homogenous nature, because they do not, (as in the *mathematical* synthesis) constitute together One magnitude, quantum. The synthesis of cause and effect, for example is dynamical."[111] Saumarez's works were dynamical in their teleology as well as in their ascription of powers to nature, and for this they merited Coleridge's praise. They were also widely ignored: Saumarez complained that no one had answered his attack on Newtonian philosophy.[112]

Coleridge, inspired by recent labors in the physical sciences and encouraged by a circle of philosophically literate friends, felt that the time had come to promulgate dynamism. He may have hoped to emulate Schelling, whose varied publications included a journal for speculative physics.[113] Certainly, Coleridge's own explorations of *Naturphilosophie* and German metascience and science were extensive enough, and they were complemented by a careful reading of the most sober English texts. His ideas about scientific method, and his schemes of the sciences, were to be well founded.

4

COLERIDGE AND METASCIENCE

APPROACHES TO NATURE AND SCHEMES OF THE SCIENCES

Unity in multeity: the need for a method

In 1803, Coleridge jotted down a note for a "Poem on Spirit – or on Spinoza – I would make a pilgrimage to the Deserts of Arabia to find the man who could make understand how the *one can be many*! Eternal universal mystery! It seems as if it were impossible; yet it *is* – & it is every where!"[1] Multeity was everywhere. The empiricist tradition in science emphasized it – indeed, for Coleridge, presented nature as made up exclusively of a world of little things. This was intellectually and emotionally unsatisfying. "I can contemplate nothing but parts, & parts are all *little* – ! – My mind feels as if it ached to behold & know something *great* – something *one* & *indivisible* – and it is only in the faith of this that rocks or waterfalls, mountains or caverns give me the sense of sublimity or majesty! – But in this faith *all things* counterfeit infinity!"[2]

The observer, faced by the vast richness of the world of the manifold, sought and needed to appreciate its minute beauties while grasping its unity. At one level, this synthetic function was carried out by the understanding, which brought together a multitude of impressions into a unity by identifying something common to them all.[3] But this was limited by the extent of the observer's experience. The understanding was not enough, and a different approach was needed.

One answer was provided by faith, through which "*all things* counterfeit infinity": "Are we struck at beholding the cope of heaven imaged in a dew-drop? The least of the animalcula to which that dew drop is an ocean presents an infinite problem, of which the omnipresent is the only solution . . . even the philosophy of nature can remain philosophy only by rising above nature."[4] Coleridge's appeal to something above nature was to God-given reason, informing nature. All nature was a single rationally and divinely created system, at least partially intelligible to man because man had his reason in God's reason. Creation, and the subsequent operation of the laws of nature, proceeded according to divine ideas. The Aristotelian axiom, that "in every true whole, the whole is prior to its parts," was thus necessarily true. Coleridge illustrated his argument with the construction of a circle: "It is . . . evident that the circle-line must have pre-existed, in its idea, in the mind of the describer, or no reason could be assigned why these points, indifferent to all figure, had not constituted a triangle, or square."[5]

It followed that no part was intelligible until one had first understood the whole of which it was a part. No least product of nature could be understood until one had understood the whole of nature. At the level of reason alone, the doctrine of ideas might work well enough, corresponding to Schelling's *Naturphilosophie* in telling us nothing directly about individual phenomena. Coleridge's love of the minute beauties of nature did not permit him to be so cavalier. His was the problem of bringing together the most exact observation and an all-encompassing structure of ideas. Faced with the endless range of natural phenomena and remaining faithful to them while seeking their underlying unity – unity in multeity, "how the *one can be many*!" – he needed and devised his own method.[6]

Method was vitally important for Coleridge, who called it "a distinct science, the immediate offspring of philosophy, and the link or *mordant* by which philosophy becomes scientific and the sciences philosophical."[7] His own method grew from an understanding of the development of the sciences, and was used to extend his grasp of them. It proved invaluable as a critical tool for assailing the unmethodical sciences of his day – for so Coleridge saw them – and repeatedly enabled him to ask crucial and seemingly prescient questions at the threshold of knowledge. It was also a fruitful and original fusion of elements of conventionally distinct approaches, the empirical and the Platonic.

Coleridge and empiricism

The observation of nature

Francis Bacon claimed in his *Novum Organum* that he had "established for ever a true and lawful marriage between the empirical and rational faculty."[8] Coleridge, in his philosophical lectures of 1818–19, announced that Christianity embraced Platonism and the common sense of experimental science.[9] Bacon had combined what he saw as best in the Aristotelian and Platonic traditions. Coleridge's professed Platonism was heavily imbued with Christian Neoplatonism. This sharing of traditions enabled Coleridge to use Bacon, "the British Plato," in a work on principles that would be valid in science and in religion.[10]

Coleridge's resolution of the problem of combining reason and empiricism in science drew on long-established traditions.[11] The observation of phenomena forms an essential part of the study of nature; Coleridge stressed that it needed to be complemented by reason and imagination.

In the *Theory of Life*, Coleridge presented man as nature's crowning achievement. Greatness in individual men was the product of balance; intellectual height, or genius, had to be balanced by a corresponding and opposing awareness of the real world and sympathy with nature.[12] Whoever was most truly intellectually or rationally alive would necessarily be most intimately attuned to nature. The interdependence of subject and object, of mind and nature, was here presented in a warmer guise than metaphysics.

Shakespeare was for Coleridge the greatest literary genius, like nature "inexhaustible in diverse powers," and "equally inexhaustible in forms." He was indeed "a nature humanized." The essence of his genius lay precisely in his sympathy with nature. Coleridge found in him the ability not merely to appreciate nature, but to become one with it. Nature was truly realized by Shakespeare in his poetry, through a love that embraced the grand sweep and form of creation, as well as the individual and the special, even the "very minutest beauties of the external world."[13]

Coleridge might have written as much of himself. He certainly possessed faithfulness to the world of nature, and sympathy with it. His love of nature and his thinking and feeling prompted by the observation of nature were mainly devoted to the minutest beauties, to a single moss or flower rather than to the starry universe. In keeping with this preferred perspective, his conversation poems, "Frost at Midnight," "This Lime-tree Bower My Prison," and the rest, contain some of his finest writing. Kant's *Theorie des Himmels*, in contrast, failed to induce in him wonder at the majesty of the heavens; there was little sublimity, he thought, in an endlessly repeated image of "a blind Mare going round and round in a Mill!"[14]

His delight was in the immediacy of careful observation, and also in its purity. He was scrupulous in his own observations and in his regard for those of others. He described minutely the structure of a candle flame, refined an observation in Gilbert White's *Selborne*, and valued Charles Hatchett's chemical analyses most highly on account of their accuracy.[15] These were all pure observations, uncontaminated by speculation, and so potentially valuable in science: "It is of the highest importance in all departments of knowledge to keep the Speculative distince from the Empirical. As long as they run parallel, they are of the greatest service to each other: they never meet but to cut and cross."[16]

He recognized distinctions within empiricism, following Bacon's *Novum Organum* in viewing experiment as a method of inquiry, a form of directed observation that went beyond mere experience. He even proposed that facts, the objective pole in his polarity between subject and object, man and nature, could be classified in ways that characterized different sciences. Physics, he asserted in 1818, was founded on facts of observation, whereas chemistry was founded on facts of experiment.[17] He had not always used this distinction. In 1801 he was exceedingly delighted with the beauty and neatness of Newton's *experiments* in the *Opticks*. But however beautiful Newton's experiments, Coleridge, as we have seen, rejected Newton's theories and his system.[18] *Natura naturata* without *natura naturans* was barren, and no sufficient source for science. Facts needed interpretation; the study of nature had to be based on the recognition of the role of the active mind.

The poverty of mere empiricism

Passivity of mind was a major deficiency not only in the end products of scientific investigation – systems, theories, and laws as proposed by contemporary scientists – but also at every step in the construction of these products. Mere observation was undiscriminating; the essence of scientific activity lay in the recognition and selection of significant phenomena. Experience was just one step above observation, raised there "by speculative philosophy . . . and not yet experimentative." It was therefore imperfect as an instrument of science. Coleridge gave, as an instance of this imperfection, the "Egyptian Naturalists" who, "having their attention roused by the swarming mud deposited during the Overflow of the Nile believed themselves to have s~~aw~~een Frogs, whose extremities (fore and hind legs) were complete, while the head and trunk were still a lump of mud . . . yet waiting for the vis plastica."[19]

Any science that pretended to limit itself to empiricism either was fraudulent in this pretension or else was no true science. Coleridge knew many surgeons and physicians and, partly through his collaborative association with Joseph Henry Green, was particularly concerned about the state of the medical profession. He manifested his concern typically through an analysis of the lessons of history, discerning principles and applying them to contemporary issues of theory and practice. Coleridge's method, here as in *Church and State*, was to arrive at the guiding idea: "I should wish to be able to address a long essay or series of Essays to the Medical Profession, pointing out in the spirit of Love their aberrations from the *Idea* of their profession, the probable causes, & the remedies."[20] The principal aberration was toward an exclusive empiricism:

> Hippocrates taught, and in a certain sense, with great truth that the Art of Healing must be learnt from the practice of Healing. Later Physicians added the word, *exclusively*; ~~and~~ the sect of Empirics commenced: and . . . it was assumed as a fundamental truth, that in order to the just discharge of medical duties the one ⟨and only⟩ thing necessary ~~was~~ in each and every case was, that the Physician should recollect from his own experience or that of others what had been successfully tried in former cases of the same kind and species. And doubtless, if all cases of Disease~~s~~ could be ⟨as⟩ safely reduced to kinds and species, as the metals or even as plants; if the differences occurring in individuals were as slight and non-essential as are observed in ⟨different⟩ specimens of the same Ore; and if, lastly, Medicine were an insulated Trade, unconnected with physiology and ~~the~~ liberal ~~knowledge;~~ ⟨science;⟩ this mode of procedure might be tolerably adequate to ~~the~~ ⟨its'⟩ immediate Objects: ⟨until a new genus or species should start up.⟩[21]

The last two conditions are the most significant ones here. Exclusive empiricism requires a complete enumeration of instances, without which it fails; the discovery of anything truly new in kind, any real extension of knowledge, breaks the boundaries and destroys the adequacy of the old empirical scheme. And that scheme, even without the addition of destructive novelty, was valid only in isolation. Once connect it with truly scientific knowledge, and its limits and explanations will be seen to be wholly inadequate. Medical men who affected the modesty of mere empiricism were guilty in Coleridge's eye of "degrading the Science in order to elevate themselves," because empiricism, being self-limiting, was necessarily barren.[22]

Medicine was by no means the only target in Coleridge's campaign. We have encountered his ridicule and condemnation of the early activities of the Royal Society of London, demonstrated in what he saw as the quintessential absurdity of Robert Hooke's proposals.[23] Hooke and his fellow virtuosi, instead of questioning nature by using experiment as an organ of reason, had indulged in mindless empirical compilation. They had failed to grasp Bacon's method. Their heirs perpetuated these follies in early nineteenth-century England – so at least ran Coleridge's intemperate assessment of such contemporary studies as mineralogy:

> And can the blind adhesion to the gross misunderstanding of Bacon's Principles, or rather to the perverse substitution of a crass idly-busy empiricism, the bustling magpie-peeping passivity of the sensualised Understanding for the true *Principles* of an Experimental Philosophy – i.e. Philosophy suggesting and dictating Experiments – first uttering the WORD *from within*, . . . and then with true Pythagorean submission of Soul / *worshipping the Echo* – the response of the Gemina Natura, – can this debasing infra-plebëian Prejudice so utterly infatuate such minds, as Davy's and Woolaston's, as to make them think it possible, that the solution should ever be afforded by a cumulus, a Rubbish-Hill, of Descriptions of the appearances which this bit of mineral and of that, and of another only to the ten thousandth presents or may by fire, & chemical tests – be made to assume? – Yes! – this will take place at the same time, that the Law & immovable Theory of the Clouds, Winds, and Rains shall have been deduced from the studious perusal of the thousand Volumes Folio, containing a compilation of all the meteorological Journals kept in all the [?world.][24]

Coleridge's scorn for compilations of observations masquerading as science was one with his rejection of the domination of the senses. He perceived this as a fundamental flaw of the age. Observation dignified into experience was still based immediately on the senses, one pole only of right intellectual method. In a biting attack on utilitarian notions of education, Coleridge pilloried exclusively empirical science and mock-Baconianism:

> Education reformed. Defined as synonimous with Instruction. *Axiom of Education so defined.* Knowledge being power, those attainments, which give a man the power of doing what he wishes in order to obtain what he desires, are alone to be considered as knowledge, or to be admitted into the scheme of National Education. Subjects to be taught in the National Schools. Reading, writing, arithmetic, the mechanic arts, elements and results of physical science, but to be taught, as much as possible, empirically. For all knowledge being derived from the Senses, the closer men are kept to the fountain head, the *knowinger* they must become.[25]

Added to this argument was the complementary philosophical rejection of the adequacy of any view of the external world based on the hypotheses of an exact correspondence between that world and our direct sense impressions of it. Hartley's psychology was an example of a system based on this hypothesis, and Coleridge dismissed it for removing "all reality and immediateness of perception" and placing us "in a dream world of phantoms and spectres, the inexplicable swarm and equivocal generation of motions in our own brains."[26]

Experience simply was not enough. Contemporary scientists presented as laws generalizations based on experience. Coleridge argued that laws had to embody causal explanations, so that to describe as laws the rules of chemical combining proportions formulated by Dalton and Gay-Lussac was for him a misuse of language. He saw the consequences of such misuse as serious – in this instance, a literal perversion and debasement of science. There was a need to assert the activity of the mind in science, and to signal this assertion through a reform in scientific language and method, inextricably interconnected.[27]

But if empiricism was not enough, it nevertheless remained essential. Just as Coleridge in one direction dismissed philosophies that derived all knowledge from sense, so did he require in the other direction that any satisfactory philosophy should account for experience. The senses and empiricism were not rejected, but subordinated to mind.

Classification: an enormous nomenclature

Facts, before they can become part of natural science, need to be arranged methodically. They need to be classified. Now methodical classification may be virtually lacking, and natural history may be presented in a sequence arbitrary in all but its conformity with the predilections of the author. Buffon, one of the leading eighteenth-century authors on natural history,[28] took the view that because all classifications were arbitrary, he would consult only his own convenience and that of his readers. There would be no deeper system in his presentation of nature. Linnaeus, best known of all eighteenth-century contributors to classification, was concerned with ease of reference and the avoidance of ambiguity. His system is precise, its rules are clear, and its distinctions are based on the generally readily observable sexual characteristics of plants. Cuvier, the leading naturalist and zoologist in early nineteenth-century France, recognized that the complexity of the world of nature invalidated any classification based on a single characteristic. He therefore combined two methods, one based upon the similarities and convergence of species, the other based upon the principle of the subordination of characteristics. This enabled him to classify animals according to what he saw as their totality and their essential nature.[29]

Clearly, for Coleridge, Buffon's approach was undeserving of serious consideration. And indeed there is no mention of Buffon in Coleridge's most extended discussion of classification, in *The Friend*.[30] In contrast, Coleridge acknowledged and even revered Linnaeus for his seminal contributions to botany: "As for the study of the ancients, so of the works of nature, an accidence and a dictionary are the first and indispensible requisites: and to the illustrious Swede, Botany is indebted for both."[31] But Coleridge complained that Linnaeus had used exclusively the external characteristics of vegetable sexuality without comprehending the place of vegetation in nature or the essence and "inner necessity" of sex. And so his system broke down – for example, with mushrooms and other cryptogams. Coleridge, always given to probing the obscurities and weaknesses of any theory as the most fruitful path to seeing how and where the theory could legitimately be developed, was particularly interested in this problematic area of botany; he made notes, for example, on C. F. B. de Mirbel's study of cryptogamous and agamous vegetation.[32] Such problems meant for Coleridge that

> after all that was effected by Linnæus himself, not to mention the labours of Cæsalpinus, Ray, Gesner, Tournefort, and the other heroes who preceded the general adoption of the sexual system, as the basis of artificial arrangement – after all the successive toils and enterprizes of HEDWIG, JUSSIEU, MIRBEL, SMITH, KNIGHT, ELLIS, &c. &c. – what is BOTANY at this present hour? Little more than an enormous nomenclature; a huge catalogue, *bien arrangé*, yearly and monthly augmented, in various editions, each with its own scheme of technical memory and its own conveniences of reference! . . . The terms system, method, science, are mere improprieties of courtesy, when applied to a mass enlarging by endless appositions, but without a nerve that oscillates, or a pulse that throbs, in sign of *growth* or inward sympathy.[33]

Linnaean botany and natural history viewed in this light were simply pointless, like an account of a military campaign that described what an outsider could see and hear while omitting any attempt to understand strategy and tactics. What was even worse was the way that artificial classifications stemming from experience and convenience could come to seem more important than the study of nature. If there was no clear place for a species in an artificial scheme, one would be tempted to ignore it.[34] Coleridge admitted that if one had not yet arrived at a scientific understanding of a subject, one should be satisfied with the forms of classification – "yet nothing less than the proof that the difficulties [of attaining a scientific understanding] are insuperable ought to preclude the attempt, or habituate us to regard it as desperate & visionary."[35]

No scheme of classification based upon a single characteristic or a single form of abstraction could hope to serve all purposes. Linnaeus's sexual scheme and schemes based upon the great chain of being failed equally in this respect.[36] Coleridge's views here coincided with Cuvier's,[37] and in doing so they merely reflected the current state of the art. What makes his discussion of classification of interest is the way he seeks to incorporate it into a philosophy of nature, and make it part of his own thought. If one was reduced to relying on artificial classification, then one needed different schemes for different purposes. But "to arrive at the ends of philosophic Arrangement, there seems to be but one way – namely, to discover and bring together before the mind the . . . principal objects which Nature effects in the process of elevating matter into organization and *manifest* vitality – and under each of these to arrange the facts illustrative of the particular idea, as far as it extends."[38] Facts needed to be arranged to illustrate ideas and principles, and thus to illuminate the constitution and not merely the order of nature. One example of such an idea advanced by Coleridge was that "the harmony between the vegetable and animal world, was not a harmony of resemblance, but of contrast," so that the lower animals were closer to the lower plants than to the higher plants.[39] The development of this idea might, he believed, inform botany and help to elevate it into what he meant by a true science, one in which ideas led one to understand truly causal laws. What was crucial was the harmony between mind and nature, the dynamic reconciliation of subject and object. Plato and Bacon had started from mind and nature, respectively. But Coleridge believed that one should be able to deduce "the Practice of Bacon from the Metaphysics of Plato." He insisted "that the Logic of both in the strict sense of the word is the same, both recommending a copious Induction of particulars & a consequent investigation of their analogies & their points of difference, & an ascending Series of Classifications built on this investigation."[40]

Classification, resting on the study of particulars and informed by ideas, could become methodical and scientific. And for this, in many branches of the study of nature, experience needed to give way to experiment guided by the active mind.

Observation, experiment, and the active mind

To Coleridge, every perception involved a creative act of mind. "The primary IMAGINATION I hold to be the living Power and prime Agent of all human Perception, and as a repetition in the finite mind of the eternal act of creation in the infinite I AM."[41] To be incorporated in science, perceptions needed to be organized by ideas and, at a very elementary level, to be interpreted. Coleridge more than once made this point by referring to the use of a microscope. One needed to learn to use a microscope in order to see with it and not just through it. Above all, one needed to learn what was significant and what should be ignored by being abstracted from consideration. Coleridge suggested a comparison between "a Metaphysician and for yet undisciplined Readers a Logician" and an "eminent Microscopic Experimenter . . . [I]n order to distinguish we must abstract, and remove; . . . the preparatory process is principally abnegative, and the final Product is a *mental* result."[42] Creativity in science involves a way of seeing that grasps essentials, and the ability to seize upon crucial phenomena that reveal ideas. The investigator's mind questions nature, and the same mind has to interpret the answers that nature gives.[43] Asking the right question is of critical importance.

Experiment, ideas, and the *prudens quaestio*

Coleridge was seeking an approach to science in which the initiative would be with mind, but in which the world of phenomena would furnish the test of theory.[44] Reliance on rationality alone would lead to "the common mistake of Entia logica for Entia realia," whereas reliance on the senses alone would leave experimentalists liable to the "Scoff of having their eyes at the tip of their Feelers, like the Snail."[45] His response was to assert that any system of science presupposed an unquestionable assumption that gave rise to an idea. The idea would generate mental constructs, consequences that should be tested, as soon as practicable, against "a parallel series of Correspondences in . . . facts of Experience" as learned from observation or experiment. The existence of correspondences between the mental constructs and facts of experience could not validate the constructs, for these had their validity from the initiating idea. But experience could demonstrate "the *universality* of the Truth, . . . that it is at once real and ideal . . . – and by this coincidence realizing the Idea and Ideal Constructions into *Laws* . . . of Nature and their genesis out of each other into powers and forces. – In the Light of Law arises Theory, and the principles of Classification – and from these again scientific Hypothesis, . . . the ~~Determination~~ (~~or~~ Proschema) of Observation, and the ~~Rationale and Direction~~ Determination of Experiment – and generally, the Architectonique and Organo~~n~~logy of Discovery and Invention."[46] His method here moves from ideas to experiments: A science barren of ideas will therefore also be barren of experiments. His strictures on botany might accordingly be supplemented by remarking that it was largely an observational rather than an experimental science, in contradistinction to vegetable physiology, in which he had a lively and sympathetic interest.[47] Only those sciences rich in ideas could progress, for "the more initiative Ideas, the greater the impulse to experiment."[48]

Francis Bacon had stressed that the "forethoughtful query" was "the prior *half* of the knowledge sought," and insisted that experiments, *planned in advance*, were decisive.[49] Coleridge made much of this aspect of Bacon, describing him as the inventor of the view of "Experiment, as an organ of reason." With Bacon, Coleridge announced, "as with us, an idea is an experiment proposed, an experiment is an idea realized."[50]

The apprehension of an idea was the fundamental act of mind for Coleridge, and the idea of polarity was central and seminal in his thought. A Coleridgean idea operates at the level of reason, beyond the limits of understanding. It can therefore not be conveyed literally, but only symbolically. Coleridge often perceived phenomena as symbols, so that, for example, he sought to convey the fundamental idea of polarity through the electrolysis of water, symbolically conceived.[51] Thus conceived, a fact became pregnant with significance, and its presentation implied more than volumes of discourse. Coleridge's illustration of the relations between ideas and experiments was of this kind, as in his appeal to the arrangement of specimens in John Hunter's physiological and pathological collections in the royal College of Surgeons. He referred to

> the life or living principle of JOHN HUNTER, the profoundest, we had almost said the only, physiological philosopher of the latter half of the preceding century. For in what other sense can *we* understand either his assertion, that this principle or agent is "independent of organization," which yet it animates, sustains, and repairs, or the purport of that magnificent commentary on his system, the Hunterian Musæum, in Lincoln's Inn Fields. The Hunterian idea of a life or vital principle, "*independent of the organization*," . . . demonstrates that John Hunter did not . . . make an hypostasis of the principles of life, as . . . a phænomenon . . . : but that herein he philosophized in the spirit of the *purest* Newtonians, who in like manner refused to hypostasize the law of gravitation into an ether, which even if its existence were conceded, would need another gravitation for itself. The Hunterian position is a genuine philosophic IDEA, the negative test of which as of *all* Ideas is, that it is equidistant from an ens logicum (= an abstraction), an ens repræsentativum (= a generalization), and an ens phantasticum (= an imaginary *thing* or phænomenon.)
>
> Is not the progressive enlargement, the boldness without temerity, of chirurgical views and chirurgical practice since Hunter's time to the present day, attributable, in almost every instance, to his substitution of what may perhaps be called *experimental Dynamic*, for the mechanical notions, or the less injurious traditional empiricism, of his predecessors?[52]

Hunter had conceived the idea of life, perhaps with less than perfect clarity. This idea had directed him in the assembly of the specimens for his museum, resulting in a literal embodiment of his idea in its progressive development through nature. Not only the overall plan of the museum, but also the selection of each specimen, was directed by this idea. Each specimen symbolized the idea in one of its aspects. And the subsequent path of experimental physiology was a pursuit of the consequence of the idea. This was the burden of Coleridge's argument. He saw Hunter's museum as exemplifying what he elsewhere called "the Method of scientific anticipation, as the Pioneer of Observation and Experiment."[53] Valid scientific method required facts of nature and ideas of reason, held in harmonious balance.

The central phenomenon

Coleridge was advocating Bacon's "true and lawful marriage between the empirical and rational faculty." But for all his high praise for Bacon, he did not follow him slavishly. Bacon had advocated the compilation of a universal natural history to furnish men of science with information. Questions could then be put to nature and answered by selective reference to this natural history – literally a history of nature. The questions he asked were designed to reveal the true nature of fundamental qualities – for example, heat. His method of enquiry was to assemble positive instances manifesting the quality or "simple nature" – the sun, fire, and boiling water, for example, all contained or manifested heat – and corresponding negative instances – the moon was in many ways comparable to the sun, but lacked heat. A careful and complete enumeration and comparison of positive and negative instances enabled Bacon to arrive at a definition of the simple nature of heat, and his definition was virtually mechanical.[54] Coleridge was seeking not such definitions, but laws, corresponding in nature to ideas of mind. These ideas could sometimes be apprehended through a central phenomenon. The idea of a central phenomenon arose from Coleridge's distinctive fusion of Platonic and empirical traditions, and its role is similar to that of Goethe's *Urphänomen*. But Coleridge's concept, empirical like Goethe's in that it refers always to a concrete instance, was perhaps even closer than the *Urphänomen* to the Platonic tradition. Coleridge was fond of dividing all men into Aristotelians and Platonists. He was much attached to the varied world of things in flux. But in science, as in many things, he believed that Aristotle and his followers had hindered the progress of physics and physiology, for all their empiricism, whereas Platonists and Pythagoreans, trusting less to experience for knowledge of nature, had yet made the greatest discoveries about external nature. Platonists had worked with ideas.[55]

Goethe had indeed written of "the pure phenomenon" as "a kind of ideal" or "idea." But it operated for him at the level of a representation, "an archetypal phenomenon in which the quality common to all instances under investigation is revealed in an unusually striking fashion."[56] Coleridge's central phenomenon was at once phenomenon and symbol of an idea, embodying a law. It was the one fact that was worth a thousand, including them all in itself, and it *made* all the others facts.[57] Each specimen in Hunter's museum needed to be apprehended in these ways. Coleridge moved from Hunter to Cuvier:

> We dare appeal to ABERNETHY, to EVERARD HOME, to HATCHETT, whose communication to Sir Everard on the egg and its analogies, in a recent paper of the latter . . . , we point out as being, in the proper sense of the term, the development of a FACT in the history of physiology, and to which we refer as exhibiting a luminous instance of what we mean by the discovery of a *central phænomenon.* To these we appeal, whether whatever is grandest in the views of CUVIER be not either a reflection of this light or a continuation of its rays, well and wisely directed through fit media to its appropriate object.[58]

Hatchett's observation was that in all ova whose embryos had bones, there was some oil, whereas there was none "in those ova whose embryos consist entirely of soft parts." Coleridge could seize on this as a central phenomenon embodying an idea because it supported his own idea of life.[59] And Cuvier's views were likewise sympathetic to Coleridge. It was Cuvier who had insisted that the anatomist "must not limit himself to a single species of organized beings, but must compare the whole; pursuing life, and the phaenomena of which it consists, throughout all animated nature." Cuvier's views on classification were, as we have indicated, close to Coleridge's. And the "Essay on method" in *The Friend* had similarities to Cuvier's statements on method.[60]

Coleridge viewed Cuvier's work as seminal because it was directed by ideas. Central phenomena were important because they too depended upon an idea in the mind. They were crucial phenomena, where the world of phenomena and the world of ideas – the latter more real for Coleridge than the former – came together. With this perspective, he decided that scientific genius lay in the ability to discover central phenomena through the apprehension of ideas in their relation to nature. "The true object of Natural Philosophy," he wrote, "is to discover a central Phænomenon in Nature; and a central Phæn. in Nature requires & supposes a Central Thought in the Mind."[61]

Ideas, nature, symbols, and correspondences

A central thought was an idea. A full account of Coleridge's use of the word "idea" would itself be another book. He wrote an elaboration that he never published.[62] He also published several concise statements:

> We assert, that the very impulse to universalize any phænomenon involves the prior assumption of some efficient law in nature, which in a thousand different forms is evermore one and the same; entire in each, yet comprehending all, and incapable of being abstracted or generalized from any number of phænomena, because it is itself presupposed in each and all as their common ground and condition: and because every definition of a genus is the adequate definition of the lowest species alone, while the efficient law must contain the ground of all in all. It is *attributed,* never *derived.* The utmost we ever venture to say is, that the falling of an apple *suggested* the law of gravitation to Sir I. Newton. Now a law and an idea are correlative terms, and differ only as object and subject, as being and truth.[63]

An idea, then, is that which, when apprehended in the knowing mind, reveals essential relations in nature. An idea thus creates unity out of multeity and shows the necessity of hitherto unperceived or uncomprehended relations. The behavior of all falling bodies was brought into a unity by the idea of gravitation in Newton's mind. Ideas also had a universalizing role, so that the behavior of all falling bodies was symbolized in a single falling apple. The tendency to individuation of all living beings was likewise symbolized in Hatchett's observation on the yolk of an egg. Coleridge had provided his own answer to his problem of 1803, "how the one can be many!"

There remained the fundamental problem of the intelligibility of nature by man: "The fact therefore, that the mind of man in its own primary and constituent forms represents the laws of nature, is a mystery which of itself should suffice to make us religious: for it is a problem of which God is the only solution, God, the one before all, and of all, and through all!"[64] Human reason and the world of nature were both created by God, and both were informed by his will and his reason. Human reason and intelligence thus had something of the divine, although in infinitely lower degree. Here was an answer going some way beyond Kant's to explain the congruence between the human mind and laws of nature. God's ideas were creative and constitutive: "A divine Idea is the Omnipresence or Omnipotence represented intelligentially in some one of the possible forms, which are the plenitude of the divine Intelligence, the Logos or substantial adequate Idea of the Supreme Mind."[65] Because ideas had in them something of divine ideas, the relation among ideas, mind, and laws followed, "there being indeed no other difference between Law & Idea than that, Will (or Power) & Intelligence being the constituents of both, in the *idea* we contemplate, the Will or power in the form of intelligence; and in the *law* we contemplate the intelligence in the form of Will or power."[66] Ideas mediated between what was real in nature and what was real in mind – they could be defined as the dynamic reconciliation of a subjectively real thesis and an objectively real antithesis. "Thus, an IDEA conceived as subsisting in an Object becomes a LAW; and a Law contemplated *subjectively* (in a mind) is an Idea."[67] Ideas existed in mind. They could lead to the highest reaches of philosophy, and to nature, "likewise a revelation of God." Nature could be studied directly, in the manner of the scientist, or read figuratively, in which case it would yield "correspondences and symbols of the spiritual world."[68]

Ideas existed in mind; yet one could apprehend them through natural phenomena that might be represented as embodying them. This apprehension could be conveyed neither literally nor directly. In the *Biographia*, Coleridge explained that "an IDEA, in the highest sense of that word, cannot be conveyed but by a *symbol*; and, except in geometry, all symbols of necessity involve an apparent contradiction."[69] Now in natural science ideas might be apprehended through Coleridge's central phenomena, which he represented, claiming Bacon's authority, as "signatures, impressions, and symbols of ideas."[70] That a phenomenon could signify a noumenal idea and unity signify multeity did indeed involve an apparent contradiction, to be resolved only by recalling that Coleridge regarded ideas as the dynamic synthesis of subjective and objective reality. Every synthesis, being between polar opposites, involved only an apparent contradiction.

His approach to nature, whether as poet or as philospher, became increasingly a quest to penetrate its symbolism. In an early notebook entry that long remained true for him, he wrote: "In looking at objects of Nature while I am thinking, as at yonder moon dim-glimmering thro' the dewy window-pane, I seem rather to be seeking, as it were *asking*, a symbolical language for something within me that already and forever exists, than observing any thing new. Even when that latter is the case, yet still I have always an obs~~cure~~cure feeling as if that new phænomenon were the dim Awaking of a forgotten or hidden Truth of my inner Nature / It is still interesting as a Word, a Symbol! It is λογος, the Creator! ⟨and the Evolver!⟩"[71] The language of nature revealed not only the laws of nature but their author. Here was an argument that Coleridge had advanced even earlier in his poetry – in "Frost at Midnight," first published in 1798, where the forms and beauties of nature appeared as

> The lovely shapes and sounds intelligible
> Of that eternal language, which thy God
> Utters, who from eternity doth teach
> Himself in all, and all things in himself.[72]

And earlier still, in "The Destiny of Nations," Coleridge had described the world of phenomena as "Symbolical, one mighty alphabet / For infant minds."[73] Recognizing and identifying nature's symbols revealed the power and creativity of the Logos, God's word. Even the words of human language were "LIVING POWERS, by which the things of most importance to mankind are actuated, combined, and humanized";[74] they had their symbolic function. How much more powerful then were the words of God's language, uttered in nature, and endowed with their own symbolism.

Coleridge was most conscious of the importance of symbols, and was careful to explain what he meant by a symbol. A symbol was a sign included in the idea which it represented; so, he argued, the instinct of ants was a symbol of human understanding, because the idea of understanding, the higher power, included the idea of the lower power of instinct, by which it was represented.[75] A fuller definition was offered in *The Statesman's Manual*: "A Symbol . . . is characterized by a translucence of the Special in the Individual or of the General in the Especial or of the Universal in the General" – that is, it represents unity in multeity, or multeity in unity. It is characterized "above all by the translucence of the Eternal through and in the Temporal. It always partakes of the Reality which it renders intelligible" – as does a central phenomenon – "and while it enunciates the whole, abides itself as a living part in that Unity, of which it is the representative."[76]

The metaphor of translucence – the act of shining through – is revealing. In a manuscript note in the copy of *Aids to Reflection* that he presented to John Hookham Frere, Coleridge distinguished between those who habitually contemplated thoughts in relation to things, and those who contemplated things in relation to thoughts. "In the former class, the Thoughts gradually sensualize: in the latter, Things light up into Symbols and become more and more intellectual. The sensual Veil of the *Phaenomenon* loses its opacity, and the *Substance*, the Numen . . . shines thro'".[77]

The representative role of a symbol meant that analogies and correspondences were its very foundation.[78] Besides German schemes, Coleridge knew one system based thoroughly and equally upon facts of science and a network of correspondences. It was the invention of Emanuel Swedenborg. Coleridge read many of Swedenborg's works with critical care, and some of the fullest statements of his own dynamic scheme of nature are in his letters to the Swedenborgian Charles Augustus Tulk.[79] Coleridge believed that Swedenborg's disciples should work toward the construction of a vocabulary and grammar of correspondences, as a step toward grasping the logical relations between phenomena and noumena; and "the Noumenon, I say, is the Logos, the *Word*" – hence the importance of high linguistic keys.[80]

The metaphor of language was pursued vigorously and variously. "The language of nature," wrote Coleridge, "is a subordinate *Logos*, that was in the beginning, and was with the thing it represented, and was the thing it represented."[81] The parallel workings of the Logos throughout the world of nature and in man – here truly the microcosm – underlay all correspondences in Swedenborg's scheme and Coleridge's. "If I mistake not," wrote Coleridge to Tulk, "one formula would comprize your philosophical faith & mine – namely, that the sensible World is but the evolution of the Truth, Love, and Life, or their opposites, in Man – and that in Nature Man beholds only (to use an Algebraic but close analogy) the integration of Products, the Differentials of which are in, and constitute, his own mind and soul – and consequently that all true science is contained in the Lore of Symbols & Correspondences."[82] Swedenborg, in his *Oeconomia Regni Animalis*, wrote of the need "to inquire what things, in a superior degree, correspond to those which are in an inferior degree," and advocated "a mathematical theory of universals" to express a science of nature proceeding from particulars to universals. Coleridge, reading this, wrote in the margin that this was "an indispensable condition of all further progress in real science."[83] He was not uncritical of Swedenborg, but judged his scheme – presented through "facts visa et audita into Symbols and Allegories" – to be grand and rational.[84] Coleridge's own schemes would be no less.

Law, cause, theory

> In order to contemplate the laws even of nature, in order to refer the phænomenon of the perishable world to a permanent law, we are constrained to consider each ~~of its~~ minute elements as a living germ in which the present involves the future & in the ~~in~~finite [?the] infinite abides potentially . . . [T]he finite can be one with the absolute, inasmuch only as it represents the absolute ~~truly~~ verily under some particu-

> lar form . . . [which] is at once the product, and the sign of the positive power [of that particular form].[85]

Coleridge's argument meant that powers in nature, the objective correlatives of ideas in mind, acted in conformity with laws of nature so as to lead to particular finite products or actualities from among the infinite range of possibilities. The potential realm was infinite, but the realm of actuality was both finite and ordered. "What is Nature? Multeity coerced into Number and Rhythm . . . How? . . . – by the WORD, and by every Word that proceedeth in and thro' the same / – Call the Words Numbers numerant, Living Numbers; or Ideas; or laws; or Spirits; or ministrant Angels; – if you please and which you please. The Terms are all equivalent."[86] They are not all equivalent, they do, however, play precisely corresponding roles in different realms – ideas in mind and spirit, laws in nature. And in reading Coleridge it sometimes seems more important to stress this correspondence than to distinguish the realms.

Laws gave the principles according to which nature was to be understood first in its forms and powers, and only afterwards by quantitative measure.[87] It was important to distinguish between true laws, which explained change through an inner necessity of things in nature, and the "⟨falsely⟩ so called Laws, the mechanical Laws" so generally subscribed to by contemporary scientists to describe what happened to bodies when they were compelled from without, regardless of their essential nature.[88]

This concept of explanation as an unfolding of the inner necessity of a development illustrates the closeness of the correspondence between ideas and laws, "Will (or Power) & Intelligence being the constituents of both."[89] That is why Coleridge often yoked reason and law together. Ideas were founded in reason as a spiritual organ. But reason was also a faculty enlightened by spirit. The structure of thought is straight out of the account of the three parts of the soul in book 4 of Plato's *Republic*. Reason operated as the "*scientific* Faculty, . . . the Intellection of the *possibility* or *essential* properties of things by means of the Laws that constitute them." In Coleridge's account, organs of sense yielded perceptions that were combined by the understanding first into individual notions and then into rules of experience. Finally, reason subordinated these notions and rules to "ABSOLUTE PRINCIPLES or necessary LAWS."[90]

The spiritual nature of reason, the relation through law of the finite to the infinite, and the role of the Logos in creating law all pointed to the same conclusion: "LAW is from God. – It is *the*

Word."[91] The unity of God's will and of the creation that obeyed it meant that there was an antecedent unity or law that rendered all parts intelligible within the whole. The antecedence was in thought, in the region of ideas, rather than in time. It was in this intellectual sense that nature was informed by the Logos and empowered by laws.[92] These distinctions between closely related ideas were important. To confuse the Logos or even laws with nature would have been to mire oneself in pantheism. This was among the dangers of natural theology. If one tried to argue from knowledge of nature to knowledge of God, there was a danger of going from theism to Spinozism. "To deduce a Deity wholly from Nature is in the result to substitute an Apotheosis of Nature for Deity."[93]

And now the distinction between nature as product and nature as productive – between *natura naturata* and *natura naturans* – must be revived. Laws governed *natura naturata*, the world of phenomenal products. But there was a sense in which they also constituted *natura naturans*, so that Coleridge's schemes of nature and of the divisions of nature may be conceived as schemes of laws and powers – hence this, among the many definitions of nature that Coleridge proposed: "The Law, or Constructive Powers, excited in Matter by the influence of God's Spirit and Logos."[94] Productivity was an essential aspect of Coleridge's concept of law: "[W]e . . . identify our being with that of the world without us, and yet place ourselves in contradistinction to that world. Least of all can this mysterious pre-disposition exist without evolving a belief that the productive power, which is in nature as nature, is essentially one (i.e. of one kind) with the intelligence, which is in the human mind above nature."[95] And elsewhere he wrote: "A productive Idea, manifesting itself and its' reality in the Product, is a Law: and where the Product is phaenomenal (i.e. an object of [?~~the~~] outward Senses) a Law of Nature. The Law is *Res noumenon*; the thing is *Res phaenomenon*."[96] Nature was thus constituted by laws.[97] Coleridge's usage differed greatly from that of most contemporary chemists and physicists. Over and over again he asserted that the law of a thing constituted its being and was the ground of its reality. "A law = the principle of the Necessity of that, which appertains to the existence of a thing . . . The principle by which it *exists* determines its *actions*: or it acts according to the principle which necessitates its mode of existence . . . Law therefore is that which necessarily determines the mode of existence of any thing, and the form, degree, and direction of its actions & relations. – In all laws a sense of necessity inheres – or it would not be.[98]

Because laws of nature were constitutive; because they were spiritual and universal, aiding man in gaining self-knowledge and knowledge of nature; and because they brought man to consider the relations between God and his creation, and so effected the spiritualization of natural philosophy, which was one pole of Coleridge's fundamental science – for all these reasons, laws were central and vital in Coleridge's thought.[99] In his unpublished *Opus Maximum* he announced his goal: "We shall deem ourselves amply rewarded for our toils . . . if at the close of the system . . . we shall have been able to refer each class of phænomena to a law præ-established ab ante and thus have demonstrated the possibility of organizing all the sciences into a living and growing body of knowledge."[100]

Just as nature was constituted by law, and was necessitated as *natura naturata*, so it could be regarded as the sum of all necessitated things, the sum of all things "comprised in the Chain and Mechanism of Cause and Effect." Here was a clear distinction between *natura naturata* and mind, the former, unlike the latter, being representable in the forms of space and time.[101] The language in which Coleridge discussed causality was generally Kantian. Coleridge's discussion of teleology was indebted to the *Critique of Teleological Judgement*.[102] He also followed Kant, whom he regarded as a supreme logician rather than metaphysician, in regarding causality as a category of the human mind. Coleridge, however, attributed to Bacon a prior statement of the discovery that "the notion of cause and effect belongs to Logic – to the arrangement of our thoughts, and dare not be supposed in nature, or rather cannot without contradiction in terms."[103] He certainly did not find it in Bacon in so many words. The vocabulary is Kant's and Coleridge knew very well the table of categories in the *Critique of Pure Reason*. He stressed that the relation of causality belonged to logic – to the understanding rather than to the reason.[104]

He was unwilling that the concept of causality should be made to descend to the level of the senses by a misinterpretation of "a constant precedence into positive causation." An example of what he regarded as a false and degraded concept of causation was Hume's account of it as derived merely from constant conjunction. Coleridge believed that the problem of cause was not so facilely resolved – it was rather one of the real problems of philosophy, whose resolution lay in the relation between the concepts of law and cause, and was thus of central importance to science.[105]

A truly scientific explanation, for Coleridge as for Aristotle, was one in terms of causes. Coleridge's views bear striking similarities to Aristotle's in this passage from the *Posterior Analytics*: "We suppose ourselves to possess unqualified scientific knowledge of a thing, as opposed to knowing it in the accidental way in which the sophist knows, when we think that we know the cause on which the fact depends, as the cause of that fact and of no other, and, further, that the fact could not be other than it is."[106] Coleridge's account of scientific knowledge has grafted the thought of Aristotle and Kant – both metaphysicians whom Coleridge valued primarily as logicians – onto Plato's account of the divisions of the soul. The graft is distinctive, and so is its fruit, Coleridge's detailed account of the order of nature.

Science aimed at the apprehension of the relation of law. But Coleridge recognized that some sciences, as yet insufficiently advanced, had to operate at least temporarily with the lower relation of theory. In a theory, "the existing forms and qualities of objects, discovered by observation or experiment, suggest a given arrangement of many under one point of view: and this not merely or principally in order to facilitate the remembrance, recollection, or communication of the same; but for the purposes of understanding, and in most instances of controlling, them. In other words, all THEORY supposes the general idea of cause and effect. The scientific arts of Medicine, Chemistry, and of Physiology in general, are examples of a method hitherto founded on this second sort of relation."[107] Coleridge's own *Essay on Scrofula* sought to subordinate a variety of observations to a theory of the disease that would have some explanatory force and would go beyond the classification of external symptoms. It would offer more than a technical aid to the memory.[108]

It might seem that Coleridge's concept of theory, assuming cause and effect, subsuming multeity under unity, and offering an explanation, was not very different from his concept of law. But the former, unlike the latter, had no necessity. Theory was simply a unified collected view of what was already known. It might have heuristic value, giving a clear view of a subject and suggesting avenues for further investigation. Unlike a law, however, it lacked seminal power – "it cannot invent or discover." It was, moreover, falsifiable: "To suppose that, in our present exceedingly imperfect acquaintance with the facts, any theory in chemistry or geology is altogether accurate, is absurd: – it cannot be true."[109] A single factual counter-instance would falsify the theory, "for there a single just exception destroys at once ten thousand apparent confirmations." Coleridge adduced Lavoisier's theory of the nature of acidity – one of the parts of the new French chemistry that Davy had most vigorously assailed – to make his point. Lavoisier had found the same substance in a number of acids, and called it oxygen, the acid-generating principle. But the combination of hydrogen with oxygen produced water, which was not an acid. And if it were to be argued that oxygen was nevertheless essential to the formation of an acid, one would need merely to point out that hydrogen and sulphur combined to form an acid. Once this was proved, and hydrogen and sulphur were shown to contain no oxygen, the theory of acids as oxides was completely subverted. Logic had its uses, even in science.[110]

Coleridge's assertion of the vulnerability of theory to fact was closely similar to Aristotle's in the *Posterior Analytics*, and also to Cuvier's.[111] Coleridge's chain of reasoning, generally robust, might become tenuous or even tangled. But his respect for facts was absolute, and it proved invaluable in controlling his accounts of the generation of nature.

The history of nature, and the productivity of nature

Coleridge, in a notebook entry of 1819, observed that

> the proper objects of knowledge . . . which may be regarded as the Poles of true Learning, are Nature and History – or Necessity and Freedom. And these attain their highest ~~form~~ perfection, when each reveals the essential character of the other in itself and without loss of its own distinctive form. Thus Nature attains its highest significancy when she appears to us as an inner power (= vis ab intra) that coerces and subordinates to itself the outward – the conquest of Essence over Form – when by Life + ⟨superinduced⟩ Finality she reveals herself as a plastic Will, acting in time and of course ǂ finitely. Here there is Process and Succession, in each Plant and Animal as an Individual, and in the whole Planet as at once a System and a Unit – and the Knowledge of Nature becomes Natural History. History e contra had ~~its~~ her consummation, when she reveals herself to us in the form of a necessity of Nature – when ⟨she appears⟩ as a power ab extra ~~she~~ that coerces and takes up into ~~her~~ itself the inner power – the conquest of ~~ordinant~~ Form over ~~the acting~~ Essence, when the Essence is rendered instrumental, = *materia*, and the Form ordinant. – But here is *Law*, and the *Ever-present* in the moving Past, the Eternal as the Power of the Temporal – and the Historic Science becomes a higher Physiology, a transcendent Nature . . .
>
> – ǂ This maintains the necessary distinct.^{n} of N. from Deity, and consequently the reality of both. Life The Finality (or establishment of one End) as superinduced – not inherent . . . God is neither [man nor nature] but the ineffable Presupponens (and in logical usage, the Presupposition) of Both: the A and Ω the Base and the Apex of the Isosceles Triangle, Man and Nature being the sides.[112]

The knowledge of nature and of history were opposite in kind; yet they existed in a fruitful tension that brought them into a dynamic unity.[113] Nature appeared as determined by an inner power whose development was toward a goal. The unfolding of power was in one sense a living process, governed by teleology; yet both life and purpose, being created by God, were superinduced on nature. An account of the development of nature, and of its successive productions, constituted a natural history. Conversely, when the successive events of history were understood in relation to laws and principles, then the inner necessity of history appeared, and the knowledge of history became akin to that of nature. Here was a proposal for a scheme of knowledge linking man to nature, demonstrating the essential unity of knowledge, and rejecting atheism and pantheism.

The scheme was commentary, or reaction, or both at once, to the writings of Henrik Steffens. Steffens was a nature philosopher. He wrote the uncritically orthodox opening article for Schelling's journal of "speculative physics," the complement of *Naturphilosophie*, giving an account of Schelling's recent writings in the philosophy of nature. Coleridge read this, as he read most of Steffens's writings.[114] Steffens sought to subordinate his knowledge of the facts of natural history – the history of all nature – to philosophical ideas. Here was an enterprise that excited Coleridge's sympathetic interest, for he too sought a philosophically reputable account of nature, conforming to the phenomena and to reason.

Now *Naturphilosophie* was flawed for Coleridge. It made nature not only alive but even the source of life. It also made nature absolute and so self-sufficient. The productivity of nature was thus derived from an inner necessity of nature, and not by an initiating act of God's will, unless God were identified with nature. That way lay atheism, pantheism, Spinozism. Schelling in his later writings sought also to reconcile philosophy and religion, but Coleridge was unconvinced and, besides, indignant with Schelling for moving toward Roman Catholicism.[115] Coleridge was seeking to elaborate a trinitarian philosophy, and his reading of Schelling and of Steffens was rectified by the spectacles of trinitarian theology; but Steffens's breadth of natural knowledge, combined with his subscription to the dynamism of *Naturphilosophie*, made him attractive to Coleridge even though mistaken in his theology. Steffens will recur throughout the remaining chapters of this book as a stimulus for Coleridge, a fount equally of information, wisdom, and wrongheadedness. Coleridge paid tribute to him in his *Opus Maximum*, where he insisted that his own system differed from Schelling's on points that were "not only momentous but essential and the difference therefore = Diversity. Let me not however fail to declare that in the several works of H. Steffens especially in his Beiträge zur innern Naturgeschichte der Erde the spirit within me bears witness to the same spirit within him – but that in him the line of its circumvolution was begun from a false centre and with too short & undistended a compass."[116]

In his *Grundzüge der philosophischen Naturwissenschaft*, Steffens had presented nature and history as "closed totalities; the whole totality manifests itself in both." He went on to qualify this by presenting them in polar opposition, in which history was the eternal model of nature, and nature the eternal image and likeness of history.[117] Coleridge's discussion of nature and history was written in the awareness of Steffens's division of knowledge. But Steffens, in another work that Coleridge knew, proposed two principal ways of looking at nature. Either one followed nature's development from step to step, recognizing in each step only the higher power of its predecessor and thus arriving at a historical account of the evolution of nature, a natural history; or one sought to understand the inner process of this evolution, the "how" rather than the "what" and "where" of nature.[118] These twin approaches are also present in Coleridge's schemes, and they, too, exist in polar opposition whose dynamic synthesis unifies and incorporates both.

"Physiogony," the production or generation of nature, was the word that Coleridge coined to express his concept of natural history. Historical sequence and the operation of productive power were both implicit in this coinage. "⟨Physiogony or⟩ Natural History," he wrote,

> ~~is rather the History of Nature's Actions;~~ has for its subject . . . the activity of productive Powers, or the sum and series of Actions having the Facts and Phænomena of Physiography (Description or Display of Nature) and [?as] the Products . . . If natural History be . . . not a misnomer, . . . it must be either the History of Nature assumed as an Agent, or of a plurality of productive Powers ~~treated~~ considered as Agents, and which taken collectively are called Nature in the active sense – even as on either supposition the ~~sum total~~ collective Products and results within the sphere of sensible Experience are what we mean by Nature passively understood.[119]

Coleridge was thus concerned with complementary modes of studying nature, all subordinated to the unity of nature and of man with nature. The generation of nature seen through its products corresponded to a history of nature as *natura naturata*; through its productivity or powers it corresponded to *natura naturans*. The former was made up of phenomena; the latter was noumenal, and so required logical rather than historical exegesis. The distinction between the logical and historical aspects of Coleridge's accounts of natural philosophy is an important one.[120]

Nature as *natura naturans* was productive, and was to that extent alive and organic, even though life was not essentially inherent in it. It was this productivity of *natura naturans* that enabled Coleridge to use nature as virtually a synonym for life,[121] although with the qualifications needed to avoid pantheism. So it was that he could urge us to "remember, that whatever *is*, *lives*. A thing absolutely lifeless is inconceivable, except as a thought, image, or fancy, in some other being."[122]

Naturans and *naturata* made a valuable distinction, in a usage sanctioned by the schoolmen of the Middle Ages, by Schelling, and by Spinoza. In spite of his intellectual rejection of pantheism, Coleridge's attitude to Spinoza was highly ambivalent, and even sometimes thoroughly favorable.[123] But still, *naturans* and *naturata* did not convey his view of productive nature with sufficient precision, and he sought more explicit alternatives. In the *The Friend* he proposed "forma formans" and "forma formata" for the energetic and material aspects of nature. Elsewhere he described the productive and self-limiting aspects of nature – a different distinction, this, and one made earlier by Schelling – by describing the "real or objective Pole of Nature" as "Natura *se inhibens*, et uniens," and the ideal or subjective Pole as "Natura evolvens ab intra, evocans ab extra, et [?omne modo] *se exhibens*."[124]

Science in England had for over a century concerned itself almost exclusively with phenomena, with products. Coleridge, like Schelling, sought to change this emphasis and so stressed productivity from within in his philosophy of nature. He wanted to show "how things *came* to *be* so" – and so his accounts, conforming to scientific knowledge of phenomena, generally followed dynamic nature, *natura naturans*, *forma formans*.[125] Phenomena were produced by noumenal powers, the constitutive principles of activity in nature.

The most difficult problem for Coleridge was to explore how nature as productivity in general worked in and through specific powers, of which the power of life interested him most.[126] He sought the intelligible generation of "the main agents ~~of~~ in all the Phænomena of observable nature, instead of assuming them as a somewhat common to the several classes, and therefore either a part of, or mere general terms for, the Phænomena, of which they are to be the solution." He saw himself here as a "philosophic Naturalist . . . a minister of the *Logos*," whose task was to show how the particular evolved from the general.[127]

His many references to the Logos, his logical account of the relations between powers, and his stress upon production and generation all led him to seek a philosophical account of cosmogony after the creation, the infusion by God's will of the actualizing principle into potential nature; he wanted to discover how the divine idea worked in nature as law.[128] Now, besides his own accounts, and those elaborated by Schelling and his followers, there was one uniquely authoritative account of the actualization of the potential at God's fiat – the Mosaic account of creation.

Creation was through a sequence of distinct acts. The productivity of nature was similarly progressive, whether conceived historically or logically. Coleridge rejected the ascent of nature assumed in the idea of the great chain of being.[129] But his schemes of nature, whether logical and in terms of powers, or historical and in terms of things, did incorporate the concept of hierarchy. Powers were arranged in sequence, and "in the *idea* of each power, the lower derives it's *intelligibility* from the higher: and the highest must be presumed to inhere latently or potentially in the lowest, or this latter will be wholly unintelligible, inconceivable." Similarly, in the phenomenal world, "in the several Classes and orders that mark the scale of Organic Nature from the Plant to the highest order of Animals each higher implies a lower in order to it's actual *existence*."[130]

Nature for Coleridge was alive and organic insofar as it was constituted by productive power. The life of nature, however, worked in nature, into which it had been infused and where it was maintained by God's will. Powers acting in nature could be arranged in a logical hierarchy, in which the power of life with its subordinate powers was at the top. Life thus inhered potentially in all beings that were products of nature. Only those beings in which life as a power was actualized were part of the scale of "Organic Nature." There was thus a general sense in which nature could be said to be alive. The consequent analogy between productivity and life referred to the *powers* of nature. At the phenomenal level, organic nature was made up of those beings in which life as a power was active. The life *of* nature and life *in* nature were therefore distinct; and the former was not part of that hylozoism that led to atheism. Coleridge's use of the word "life," like his use of "nature," has several meanings.

This distinction is pertinent to Coleridge's account of hierarchy in nature. It was equally true of the scale of organic nature and of the logical scale of powers that the hierarchy was not a continuum: "for this is one proof of the essential vitality of nature, that she does not ascend as links in a suspended chain, but as the steps in a ladder." This account, though it conveyed an image of distinct stages in the ascent of nature, was still imperfect, for it suggested that all nature was contained in a single line of ascent. Coleridge therefore went on to suggest a more complex alternative simile: Nature "at one and the same time *ascends* as by a climax, and expands as the concentric circles on the lake from the point to which the stone in its fall had given the first impulse."[131] And this simile in turn needed elaboration and the introduction of further discriminations. Nature was characterized by a tendency to act in every part. The simile of ripples on the surface of a lake had to be interpreted differently for inorganic and organic nature. In the former, subjects were properly to be contemplated relative to the common center, the center of propagation of the ripples. In the latter, a subject had its own center; it was "idiocentric," and its development had to be contemplated relative to this center. "Still however the same Nature," Coleridge observed; "proof of it, that equally in the inorganic as in the organic World Nature acts in both ways." Indeed, there was in nature a "law of Bi-centrality – i.e. that every Whole . . . must be conceived as a possible center in itself, and at the same time as having a Center out of itself, and common to it with all other parts of the same System."[132]

Classifications derived from such a view of the development and ascent of nature thus introduced distinctions based upon relations of powers and of things. These relations existed in our minds because of a corresponding unity in nature: "We *divide* in classifying in order to be able to distinguish – but Nature produces distinction without breach of continuity." Inorganic and organic nature stood in polar opposition one to the other, but "both alike are the same."[133]

Powers and polarity

Coleridge considered that all our inquiries about nature could be generalized into two questions:

> First Question. What are the POWERS that must be assumed in order for the Thing to *be* that which it is: or What are the primary Constituent POWERS of Nature, into some modification or combination of which all other Natural *Powers* are to [be] resolved?
>
> Second Question. What are the Forms, in which these Powers *appear* or manifest themselves to our Senses[?][134]

The answer to the second question would be a synoptic key to the sciences. And Coleridge's answer to the first question was "that all the primary Powers of Nature may be reduced to Two, each of which ~~in like manner~~ produces two others, and a third as the Union of both." He went on to explain that the two primary powers were in polar opposition. Thus his account of productive nature was to be in terms of powers and polarity, whose definitions were interdependent. Powers could act and manifest themselves only by opposites, each depending for its existence on the existence of the other; "and the process itself, in which THE ONE reveals its Being in two opposite yet correlative Modes of Existence, I designate by the term, Polarity, or Polarizing."[135]

Powers and polarity between them would enable the philosopher of nature to construct an intelligible account of productive nature. This was Coleridge's claim in the opening paragraph of chapter 13 of the *Biographia Literaria*:

> DES CARTES, speaking as a naturalist and in imitation of Archimedes, said, Give me matter and motion and I will construct you the universe. We must of course understand him to have meant, I will render the construction of the universe intelligible. In the same sense the transcendental philosopher says; grant me a nature having two contrary forces, the one of which tends to expand infinitely, while the other strives to apprehend or *find* itself in this infinity, and I will cause the world of intelligences with the whole system of their representations to rise up before you. Every other science pre-supposes intelligence as already existing and complete: the philosopher contemplates it in its growth, and as it were represents its history to the mind from its birth to maturity.[136]

Coleridge was proposing an account couched in the language of dynamic logic to render intelligible the history of nature. There would be a correspondence between the logical account of noumena and the history of phenomena.

In this generation of dynamic nature, one began with a productive power that, like nature itself, was susceptible of growth. Coleridge held that any significant philosophical account of nature should exhibit the relation between unity and multeity – should show how individual things are related to particular and to universal laws, by demonstrating that "the individual must present to us some power some principle of actuality."[137] Things were produced by powers, and existed in virtue of the continuing operation of powers. Powers were constitutive actualizing principles, related to ideas and to things: "An Idea is a POWER . . . that constitutes its' own Reality – and is, in order of Thought, necessarily antecedent to the *Things*, in which it is, more or less adequately, realized."[138] And this relation was reinforced by the congruence that Coleridge assumed between laws and constructive powers.[139] Powers were accordingly symbols of the process of creativity. Moreover, the dynamism of polar opposites and their unifying reconciliation, +, −, and 0, was a trinitarian metaphor, to be applied in philosophy, science, and theology. Coleridge's dynamic logic was an instrument maintaining the unity of his thought and the congruence between his interpretations of science and theology.

If one considered an interrelated set of changing appearances within a particular system, those changes could be explained by reference to the power or law of the phenomena. The power, seen in connection with reason, was an idea; "and Philosophy is the Science of IDEAS: Science the Knowledge of Powers." Coleridge, dissatisfied with the senses as the source of a knowledge of reality, believed that it was important to distinguish the relations of things or products from the things themselves, and to seek an account of relations in terms of powers.[140]

For example, scientists had erred in their account of the metals, some, like Lavoisier, seeing them as so many different elements, others, like the *Naturphilosoph* Oken, recognizing their kinship but attributing it to a component common to them all. Coleridge dismissed this as "Alchemistische Nonsens, unworthy of Oken who ought to have known that the Radical of the Metals must be a *Dynamis*, not a Thing"; that is, their kinship resulted from common constitutive powers.[141]

Powers could be invoked to account for the qualities of bodies, such as metalleity. They were indeed the productive source of all qualities, and thence of quantity.[142] But there were not merely substantive powers in Coleridge's classification, but also modifying and ordinant ones, so that the universe of things, their changes, and their arrangements could all be referred to the activity of powers.[143] Activity is vital to the metaphor. Coleridge's powers were not Newtonian forces, acting on matter, and differing from imponderable fluids only in being immaterial vehicles for the action of phenomenal forces. Beddoes and Davy had invoked such powers or forces as alternatives to Lavoisier's fluid of caloric, the matter of heat, and Franklin's electrical fluid.[144] But for Coleridge, powers were prephenomenal, and more real than phenomena because less deceitful than them. He regarded forces as evolutes and phenomenal products of powers. Nor were powers ever passive in the way that caloric could be when combined with ponderable matter. Activity was of the very essence of power: "By generalizing a continuous Act, or a series of Acts essentially the same, and then contemplating this generality as a Unity, we form the notion of A POWER." The scheme of nature would be complete when it was understood as a scheme of productive activity in which "all the powers that now are have been duly evolved each in its own branch in the tree of philosophic genealogy, the great pedigree of the world."[145]

Once grant polarity, and Coleridge's cosmogony and physiogony followed. Polarity – "a nature having two contrary forces" – was all that he and his fellows in transcendental philosophy needed to render intelligible the construction of the world. It was the central metaphor of *Naturphilosophie*. Schelling had urged philosophers to search for polarity throughout nature, which he saw as informed and activated by two opposite principles.[146] Coleridge recognized Schelling as the immediate spur to the revival of the principle of polarity, but would not grant him priority. "Schelling," he wrote to his Swedenborgian correspondent C. A. Tulk,

> is the Head and Founder of a philosophic Sect, entitled Natur-philosophen, or Philosophers of Nature. He is beyond doubt a Man of Genius, and by the revival and more extensive application of the Law of Polarity (i.e. that every Power manifests itself by opposite Forces) and by the reduction of all Phaenomena to the three forms of Magnetism, Electricity, and constructive Galvanism, or the Powers of Length, Breadth, and Depth, his system is extremely plausible and alluring at first acquaintance. And as far as the attack on the mechanic and corpuscular Philosophy extends, his works possess a permanent value. But as a *System*, it is little more than Behmenism, translated from visions into Logic and a sort of commanding eloquence.[147]

Boehme seemed indeed to Coleridge to have enunciated the law of polarity, albeit in allegorical form.[148] But Boehme no more than Schelling impressed him as the first discoverer of the law. He believed that this title should go to Heraclitus, and that Giordano Bruno had revived the law two thousand years later and made it "the foundation both of logic, of physics, and of metaphysics." In so doing, Bruno was "in this as in many other instances anticipating the ⟨Ideas &⟩ discoveries ⟨generally attritubed to far later Philosophies, even those⟩ of the present ages." In the *Biographia*, Coleridge explained the similarities that he perceived between Schelling's education in philosophy and his own. Both had studied Kant; "we had both equal obligations to the polar logic and dynamic philosophy of Giordano Bruno"; and both had learned "affectionate reverence" for Boehme and other mystics.[149]

Traditional logic, Aristotelian logic, was a precise tool for rendering explicit knowledge already implicit in a set of propositions. It led to clarity of exposition and a careful and thorough enumeration of knowledge. In dynamic logic, a contradiction is resolved by a resolution or development that leads to new knowledge. Powers in Coleridge's logic enabled him to construct a metaphysics of quality, from which quantity was derived.[150] Here was a logic that was productive, a method of generating new knowledge, and more than incidentally an alternative to the metaphysics of quantity that dominated and continues to dominate the sciences.

Logic for Coleridge was burningly alive. He used it as an integral part of every realm of thought and knowledge. Here is polarity at work:

> Contemplate the Plants & the lower species of animal life, as Insects – then we may find at once an instance & an illustration of the poetic process. In them we find united the conquest of all the circumstances of place, soil, climate, element &c over the living power, & at the same time the victory of the living Power over these circumstances – every living object in nature exists as the reconciliation of contradictions, by the love of Balance. – The vital principle of the Plant can ~~embody, that is, can manifest itself;~~ make itself manifest only by embodying itself in the materials that immediately surround it, and in the very elements, into which it may be decomposed, bears witness of its birth place & the conditions of its outward growth – On the other hand, it takes them up into itself, forces them into parts of its' own Life, modifies & transmutes every power by which it is itself modified: & the result is a living whole, in which we may in thought & by artificial Abstraction distinguish the material ⟨Body⟩ from the indwelling Spirit, the contingent or accidental from the universal & essential, but in reality, in the thing itself, we cannot separate them.[151]

Poles are copresent, and nothing exists in one pole alone. Things, indeed, are the synthesis of opposing energies, and the "life of Nature consists in the tendency of Poles to re-unite."[152]

The law of polarity was, with the law of identity, one of the two great laws of nature. It was the foundation of polar logic, and also an idea that had to be imaginatively apprehended and created anew in every mind that would rationally perceive it. Schelling's revival of polarity had thus been a rediscovery, no matter what his sources; and so had Coleridge's. A marginal note to F. F. Runge's *Neueste Phytochemische Entdeckungen* shows that in the late 1820s Coleridge considered that he himself and Joseph Henry Green had together "evolved" the law in or around 1817.[153]

In *The Friend* of 1818, he gave his clearest statement of the law: "EVERY POWER IN NATURE AND IN SPIRIT *must evolve an opposite, as the sole means and condition of its manifestation*: AND ALL OPPOSITION IS A TENDENCY TO RE-UNION . . . The Principle may be thus expressed. The *Identity* of Thesis and Antithesis is the substance of all *Being*; their *Opposition* the condition of all *Existence*, or Being manifested; and every *Thing* or Phænomenon is the Exponent of a Synthesis as long as the opposite energies are retained in that Synthesis."[154] It was all very well to state the law: How could he aid the reader toward an apprehension of the idea? Ideas could be conveyed only through symbols and metaphors, not through definitions and the language of the understanding. The *Naturphilosophen*, following Schelling, used the magnet as a symbol for polarity, and so did Coleridge.[155] The poles of a magnet were truly opposite, yet they made the magnet a unity. They could not exist in isolation, and if a bar magnet was broken in two, the fragments appeared again as bipolar, each with newly produced opposite poles. Here were thesis and antithesis. But to illustrate synthesis, also called indifference, Coleridge turned instead to chemical combination. Water was neither oxygen nor hydrogen, nor any mixture of the two, but their synthesis in a new body: "And as long as the copula endures, by which it becomes Water, or rather which alone *is* Water, it is not less a *simple* Body than either of the imaginary Elements, improperly called its Ingredients or Components."[156] The contrast with the chemistry of atomism, corpuscular science, and mechanical philosophy could scarcely be clearer.

Thesis, antithesis, and synthesis are the triune aspects of polarity. But the polar logic is not merely polar, but productive, so that Coleridge's schemata exhibit polarity as the product and resolution of a prior identity, thesis and antithesis as the polarization of a prothesis. "Observe. Identity (or the One (Monas) containing the Many potentially, and contemplated as *anterior* to the production or evolution of the many) is either *Absolute*, and in this sense predicable of *God* alone; or *Relative*: in which sense we may consider a Germ, or Seed, as the relative Identity of a Plant." Polar logic thus enabled Coleridge to discourse with equal facility of the growth of a flower and of God's creation of the cosmos. But when he came to consider cosmogony, he began after God's initial act of will had polarized chaos, and "the law of polarity has commenced." Absolute identity was beyond the reach of science.[157]

The magnet has been presented as a symbol of polarity, and magnets exert magnetic forces. More precisely, in Coleridge's view, the power of a magnet exerts positive and negative forces. We can then say that "the Polarity of a Power implies its' Forces: and its' Poles *are* its' Forces."[158] Polarity worked in all physics, as attraction and repulsion, positive and negative magnetism, and the opposed forces of electricity.[159] But it also worked in every other realm of thought and nature. Law and religion, and church and state, were two manifestations of polarity in the nation.[160] There was even polarity in the moral law, discernible in the alternating attachments of "the public mind."[161] The metaphor was tempting in its universality. It is as well to remember that it had for Coleridge a precise meaning, conveyed in the following diagram:

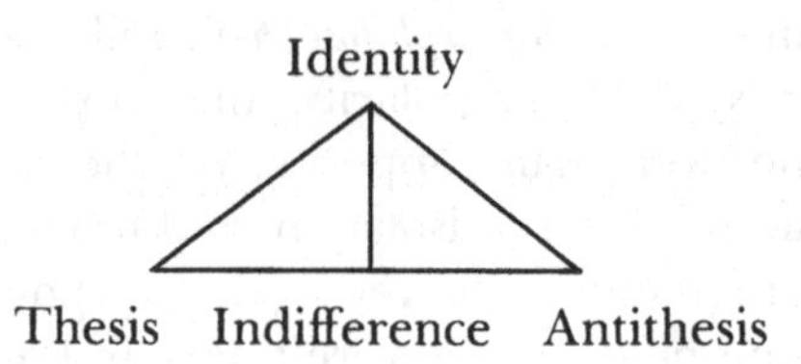

He was indignant at the misuse of the term, complaining of its Jack-of-all-trades use, as when G. A. Goldfuss, a German zoologist and *Naturphilosoph*, described carbonated and decarbonated blood as *polarisch*.[162] Polar logic was to make thought more, not less, precise, and to enable Coleridge to work toward the construction of a rigorous scheme of nature.

The compass of nature

The essential terms of Coleridge's polar logic are five: an initial *prothesis*, in which act and being are essentially one, and which generates a *thesis* and an *antithesis*. These, existing in polar opposition to one another, give rise to a *mesothesis*, or *indifference*. Finally, thesis and antithesis yielded a new *synthesis*.[163] This was the pentad that made Coleridge's logic dynamic and productive and that he applied throughout his studies of natural science and the philosophy of nature. His "great scheme or formula of all logical *Distribution* of our *Conceptions*" was adumbrated by 1815, explicitly formulated by 1817, and still central in his thought in 1830.[164] It was, in short, the organizing formula in the construction of his system of nature.

A striking instance of the application of this formula comes from a notebook entry of around 1827, in which Coleridge considered contemporary ideas about race.[165] Blumenbach had asserted that there were five principal varieties of mankind, giving them in 1795 the now familiar names of Caucasian, Ethiopian, American, Malay, and Mongolian. He had gone on to argue that the Caucasian was the original stock, from which the others had developed by degeneration.[166] This view was soon contested, notably by James Cowles Prichard, who came to Bristol as a student in 1802 and again as a physician in 1810, when he was at first associated with John King, Beddoes's former assistant. He married Anne Marie Estlin, daughter of Coleridge's old friend John Prior Estlin. Coleridge seems not to have known Prichard personally, but he at least knew of his *Researches into the Physical History of Man*, in which Prichard had argued that races arose through progressive development, through "the transmutation of the characters of the Negro into those of the European, or the evolution of white varieties in black races of men."[167] Lorenz Oken, an erratic subscriber to *Naturphilosophie* who nevertheless compiled one of the best natural histories of the early nineteenth century, had another developmental thesis, seeing four kinds of man descended from the orangutan.[168] Coleridge ignored the fourfold division and adopted Blumenbach's five varieties, arranging them according to the rules of dynamic logic:

	Prothesis Caucasian	
Thesis Negro Barbarous	Indifference Malay B	Antithesis American Savage
	Synthesis Mogul At once Barbarous [?]	

> Now it is of the utmost importance to determine previously whether the Pentad is to express progressive Development (as Dr Pritchard from the Negro, or Oken from the Oran[?utan], to the European) or Degeneracy. Now my Pentad & my Convictions are framed on the latter: & consequently I begin or find my Prothesis (the relative Identity) in the Caucasian as the least degenerate. – Still, however, tho' in reverence of my old and dear Friend, BLUMENBACH, I have adopted his ~~desig~~ names & mode of applying the theory of KANT, I prefer my own Scheme.[169]

Coleridge accordingly replaced Blumenbach's terms with Noah as prothesis, Shem and Japhet as thesis and antithesis, Ham as indifference, and the Christian, who "exists only *idealiter*," as the intended synthesis. The logical pentad, married to biblical exegesis, had given Coleridge his key to anthropology.

The pentad was susceptible not merely of application, but of development too. "The modification of Thesis by Antithesis, and *vice versa*, . . . convert the Pentad to a Heptad." And the heptad furnished Coleridge with a scheme for colors – red and "decomponible green" were prothesis and synthesis, yellow and blue were thesis and antithesis, and "indecomponible green" was the indifference; the heptad was completed by orange and indigo, the indifference between red and yellow and between red and blue. The scheme of colors was one to which Coleridge constantly returned over more than a decade, from 1818 to 1819 until at least 1832. If its details were problematic for him, its form was clear: "The true theory of Colors, as of all physics, is but a position – special of the fundamental Relation, Extension and Intension as the two polar forms of Tension (*Spannung* of the late German Philosophers, ενεργεια of the Greek. . . .)." It all came back to polarity and to the pentad.[170]

The pentad was a logical relation of five within an all-encompassing unity, "5 = 1 in all the simple proportions of the five forces of the One Power," as Coleridge wrote in August 1817.[171] It embodied four exponents of two polarities, "4 = 1 the Pan begotten by the 1 + 1 × 1 + 1," and a fifth generated by their "co-involution."[172] Generation and productivity were the keys to the comprehension and use of the pentad. Thus, for example, Coleridge represented the two polarities of positive and negative magnetism and positive and negative static electricity as generating galvanism, the power produced by a battery. In a lengthy notebook entry written in the 1820s,[173] he stressed productivity even above polarity, so that the generation of galvanism was abstractly represented in this way:

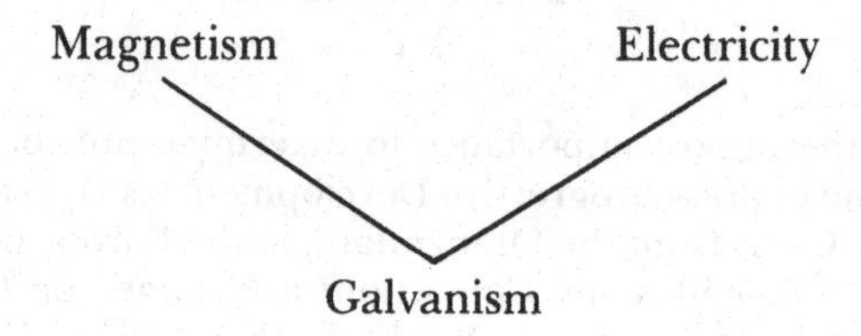

rather than as a previous formulation[174] had indicated:

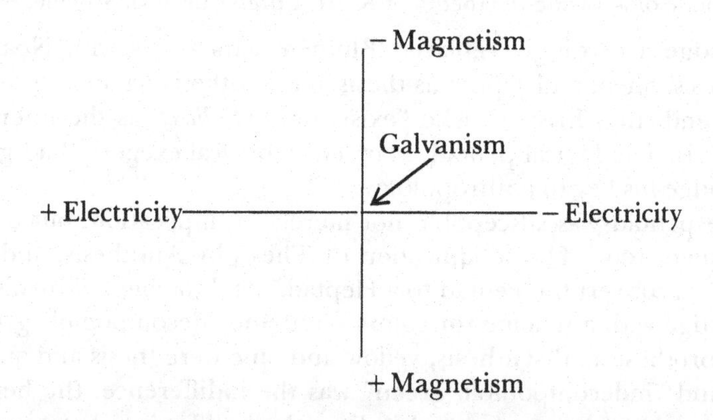

Now the emphasis upon the Trinity was his own and, never absent after the abandonment of his youthful Unitarianism, came increasingly to dominate and to inform his thought. The emphasis upon two fundamental and productive polarities came from *Naturphilosophie*. Coleridge, while not rejecting the idea of these polarities, gradually subordinated them to his logical trinity. In 1820–1 he remarked that he was "more and more inclined to prefer . . . Δ to the + of the modern Teutonic Physiosophers (= Naturphilosophen.) The Pentad would comprehend both."[175] And so it does. But the polarities of the *Naturphilosophen* remained fundamental even when subsumed under the pentad.

Schelling was responsible for developing polarities into a philosophy of nature. But it was Schelling's follower Steffens who, widely informed in the sciences, made the most varied application of "the + of the modern Teutonic Physiosophers." He gave a systematic exegesis of this quadruplicity in his *Grundzüge der philosophischen Naturwissenschaft*, a work that Coleridge knew well and first read no later than October 1815.[176]

Steffens began (p. 22) with two lines intersecting one another at right angles and representing being and becoming, respectively. He then proposed a correspondence between these lines and the dimensions of length and breadth, the first two stages in Schelling's construction of matter.[177] The third and final stage, breadth, was the indifference of the first two, and was represented by the point of intersection of the two lines:

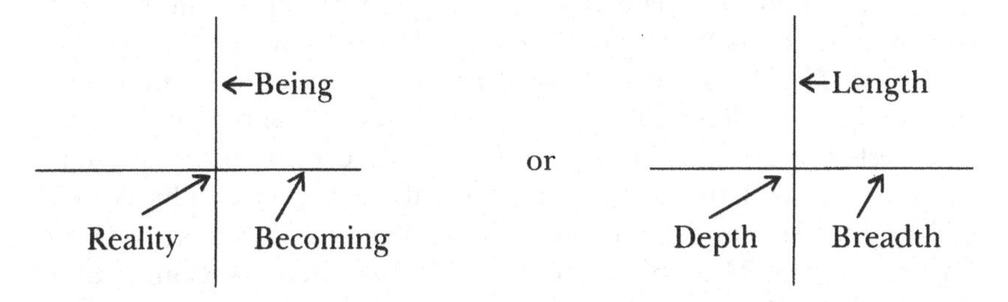

Steffens then introduced what was to be a fundamental metaphor in his and Coleridge's schemes: the compass of nature (pp. 40–1). The line of being was the north–south line, representing gravity, length, and magnetism, whose forces were exponents of the power of length. The line of becoming was the west–east line, representing light and electricity, whose forces were exponents of the power of breadth; that was why an electric charge distributed itself over the surface of a conductor. The cardinal points on the compass, north, south, west, and east, corresponded to powers, to forces, and to phenomena. And because this was a universal logic, the four cardinal points corresponded in every realm of nature to the four fundamental powers and forces, and to the phenomena symbolizing them. North, south, west, east; earth, air, water, fire; positive and negative electricity; coherence, incoherence, expansion, attraction; carbon, nitrogen, hydrogen, oxygen – these were among the correspondences, so that the north pole of the compass of nature represented earth, positive magnetism, coherence, and carbon, at the geological, ideal, potential, and chemical levels in the hierarchy of nature (pp. 40–8). The point of indifference, the intersection of the north–south and west–east lines, was not given great importance by Steffens, but he did identify chemical processes with the process of indifferentiation. Now Coleridge set great store by the fifth power, centrality, as a distinguishing feature of his system, for it was this that elevated it from a scheme of quadruplicities to a system based upon the dynamic logical pentad, with its implicit trinity.[178] This difference was ultimately far more significant than any differences in detail between his and Steffens's series of correspondences. The equation that Steffens drew between indifferentiation and chemical process must have added an intellectual reason to the biographical accidents that had led Coleridge to rate chemistry so highly. It was to be truly central in his scheme of the sciences.

Steffens, then, did not attach great significance to the fifth power. This led him to wander away from dynamism and sometimes to regard the relation between the poles and the center as one between constituents and compound, as in corpuscular chemistry.[179]

Steffens, Coleridge argued, was guilty of ignoring the "essential diversity" that he and Schelling had pointed out between the north–south and west–east lines, the former representing constituent or substantive powers, the latter representing modifying powers: "The Powers of Nature, her Hand, as it were, form a Pentad – 5 fingers, or four fingers and the Thumb. 1. Attraction. 2. Repulsion. 3. Contraction. 4. Dilation. 5. Centrality – Let the Thumb represent the 5th or Central Power. – Of these the 1st ~~and~~2nd ~~are~~ and 5th are the Constituent, *substantive* Powers: the 3rd and 4th the modifying or adjective Powers: 1 and 2 = Magnetism; 5th = Galvanism; 3 and 4 = Electricity."[180]

This and similar schemes may seem arbitrary and airy word spinning, of no conceivable interest to the student of nature. Coleridge, however, believed that a proper grasp of the pentad of powers and the compass of nature was a vital step, perhaps the crucial one, in understanding the genesis, significance, and interrelations of natural phenomena. Once understand the compass, "and you will hear the voice of infant Nature – i.e. you will understand the rudimental products and elementary Powers and Constructions of the phaenomenal World."[181] It was correspondingly important to be accurate in ascribing powers to their respective poles, to discriminate between the different kinds of fundamental powers – substantive and modifying – and to understand the possible relations between the powers. Coleridge remarked on "the urgent importance of *auseinander*ing the twofold possible relations of each of the 5 elementary Powers to each one of the remaining."[182] These interrelations were the very web of nature, constituting its unity.

Here, with a new logic, was the ancient "Heraclitic Physics" of a universe of perpetual flux, a world far removed from the fundamental material stasis of the physics and chemistry of the age. The *Naturphilosophen* were, however imperfectly, rediscovering the universe of change that the alchemists had known before them. Coleridge proceeded to argue that the mercury and sulphur of the alchemists were their symbols for light (east–west) and gravitation (north–south), and gold the culmination and interpenetration of four powers.[183]

He believed that Heraclitus, the Pythagoreans, the alchemists, Giordano Bruno, Jacob Boehme, and the *Naturphilosophen* all had some true insights into the order of nature. His own dynamic logic and his knowledge of natural science would combine all that was valid in their insights and would produce a true system of nature. The system was to be built on a series of symbols and correspondences, to that extent representing an enterprise akin to Swedenborg's. Coleridge developed his own set of symbols for powers, relations, and substances and embodying powers. It is significant that they are tabulated and explained on the flyleaf of Swedenborg's *De Equo Albo.*[184]

Genetic schemes of nature and hierarchies of the sciences

Coleridge's system was to be logical, and yet it would elucidate and recapitulate the history of nature "assumed as an Agent, or of a plurality of productive Powers . . . considered as Agents."[185]

In a particularly optimistic note of 1815, which formed a draft for part of chapter 12 of the *Biographia*, and in which Coleridge was closer to Schelling than he was later, Coleridge made explicit the general relations among nature, history, and logic. Powers of mind and powers in nature were in a rich correspondence that, once confirmed, would validate the schemes of powers:

> Our Business then is to construct a priori, as in Geometry, intuitively from the progressive Schemes that must necessarily result from such a Power with such Forces, till we arrive at Human Intelligence, and prospectively at whatever excellence of the same power can by human Intelligence be schematized. – Then will there arise a confirmation of the Truth of the Process, should it appear that all the different Steps of the

> Process, which we had shown to be the necessary Preconditions of Human Intelligence did actually exist in Nature, & that in giving the hypothetical Progression of our Self to Reason, & Conscience, we had undesignedly given the History of the Material World, from the lifeless clod to the half-reasoning Elephant.[186]

A good part of the *Opus Maximum* is devoted to this logical and historical cosmogony. The compass of nature and polar logic led Coleridge to two forms of schemes of powers, one close to transcendental philosophy and *Naturphilosophie*, the other grown beyond it and essentially trinitarian in structure.

In the first form, illustrated in a notebook entry probably dating from 1815,[187] the pentadic structure of the compass of nature provides the motif, repeated at each level of the ascending hierarchy of powers and sciences. Each level is distinct, although corresponding to every other level in the disposition of its powers. Steffens's *Grundzüge* and *Beyträge*, bound together as a single volume on Coleridge's instructions, were the models, but the nomenclature exhibits some significant individuality. The explicit identification of the first level as Ideal, asserting the logical primacy of the ideal over the real, may indicate Coleridge's rejection of nature as absolute. The Cosmical level, coming next, exhibits the powers active in cosmogony. Next, and self-explanatory, come the geological powers. The Potential level is intermediate between the geological and chemical levels, and its powers are those most directly symbolized by the cardinal chemical substances. The chemical level is in turn succeeded by the levels of organic life, of which the first is the vital level where the powers are arranged according to the infusion of *Naturphilosophie* with Brunonian physiology.[188] Finally, in the higher realms of physiology and comparative anatomy, comes Coleridge's Organic level, where the powers correspond to systems within the body. This scheme was among his earliest ones, but in its powers and sequence, though not in its arrangement, it had much that remained in later formulations. It also reflected with fair accuracy the order of Coleridge's subsequent study of the sciences, as far as this was systematic, and so it can be viewed as a plan of study as well as a scheme of knowledge, powers, and the history of nature.

Now the pentad, as we have seen, includes both tetrad and triad. If one takes any level in the hierarchy, the cardinal points constitute the tetrad, whereas the north–south and west–east lines, each taken as a unity – say, at the potential level, magnetism and electricity – together with the power represented by their intersection, which in this case would be galvanism, constitute a triad.[189] And indeed the earlier levels in the hierarchy are here presented as pentads, whereas the vital level is presented as a triad. Coleridge, in gradually dissociating himself from *Naturphilosophie*, came increasingly to emphasize the triad. In the *Opus Maximum* he quoted appreciatively from Baxter's *Life*: "The Divine Trinity in Unity hath exprest itself in the whole frame of Nature and Morality."[190] And in a notebook entry of the 1820s[191] he presented the "Genesis and ascending Scale of physical Powers, abstractly contemplated," exhibited in a series of twelve triads. These represent the successive stages in the construction of a scheme of nature, starting with existence as the synthesis of "Power" and "Sphere," perhaps an assertion within the broad tradition of Newtonian dynamism that matter exists by virtue of its powers. Then come force, motion, extension, depth, the "globific" power (akin to gravitation), integral bodies, galvanism, products of combustion, sensibility, the power of animal life, and the conscious self, as syntheses in each of the ascending triads, respectively.

The scheme draws on a wide variety of sources. Kant's *Metaphysische Anfangsgründe der Naturwissenschaft* (Riga, 1786), for example, helped Coleridge to derive motion from existence in space and time;[192] depth, or the inward power of bodies, was derived in accordance with Schelling's construction of matter, and so was the synthesis of galvanism.[193]

Coleridge viewed his scheme as one that would "⟨call forth in the student⟩ the faculty of recognizing the same Idea or radical Thought in a number of Things and Terms which he had ⟨never⟩ previously considered as having any affinity or connection." This would be achieved first through the universality of the genetic triad of thesis, antithesis, and synthesis, which underlay the order of nature, and secondly through the identification of particular phenomena with the appropriate triad. The scheme of triads, like the compass of nature, helped one to look at things and to perceive through them the underlying and unifying activity of powers. "You have but to remember, that what in one dignity . . . is a *constituent* power, in the next dignity is the *Product*, or *thing* constituted. . . .; and in the ~~next~~ 3rd, it is (= manifests itself as) the *property*, active quality or *function* of that Thing."[194]

Coleridge's schemes were not merely for the sequential construction of nature; they were also vital aids to the imaginative apprehension of the active unity of Nature.

5

THE CONSTRUCTION OF THE WORLD

GENESIS, COSMOLOGY, AND GENERAL PHYSICS

The construction of matter

Length, breadth, and depth are the three dimensions of matter corresponding to the linear axis of magnetism, the superficial charge given to bodies by static electricity, and the inner chemical effects of a galvanic cell. They can be seen as the keys to the creation of three-dimensional matter and of the cosmos, whereas the activity of the powers of attraction and repulsion, symbolized in the magnet and corresponding to the power of length, represents the first stage of that construction. Because the creation of matter and of the cosmos was initiated and effected by an act of divine will, matter could provide an image of that act of will, reminding one of the spiritual origin of nature. Thus Coleridge once described matter as "the phantom of the Absolute Will."[1] To him, matter was at once a reality and an abstraction, to be distinguished from body: "When we *think* Body in the abstract, we call the result Matter: when we *imagine* Matter in the concrete, we call it Body." His concept was here analogous to Aristotle's, wherein matter and form underlay the actuality of bodies. But for Coleridge, unlike Aristotle, matter could exist independently of body – for example, in the phenomenon of light, which he regarded as material although not corporeal.[2]

Matter, then, was generated by an interaction of polar powers, existed in virtue of those powers, and was essentially active. Coleridge, in arriving at this view, was drawing on Newtonian traditions and on the writings of Kant, Schelling, and Steffens.

Isaac Newton, in the thirty-first query to his *Opticks* (1706), had postulated atoms endowed with different powers to account for the various modes of activity of matter. Gravity, magnetism, electricity, chemical affinity, cohesion, and other phenomena could then be explained by assuming gravitational, magnetic, and other forces exerted between atoms. The general scholium to the *Principia* also discoursed about active principles in nature. Many eighteenth-century Newtonians stressed this activity while giving less emphasis to the role of atoms in nature. Gravitational attraction was exerted throughout the cosmos, whereas the bodies on which it acted occupied an increasingly insignificant proportion of space. Perhaps all of corpuscular nature might be contained in the space of a nutshell. Or perhaps, as Joseph Priestley argued in his *Disquisitions on Matter and Spirit,* there was no corpuscular matter, but only forces that constituted all matter and rendered it, like spirit, essentially active.[3]

Immanuel Kant was a Newtonian. Newton's physics and his own pietist Christianity were two certainties in his cosmos. He had studied mathematics and philosophy at the University of Königsberg; his first lectures as *Privatdozent* were on philosophy, mathematics, and physics. At the same time, in 1755, he published his *Allgemeine Naturgeschichte und Theorie des Himmels, . . . nach Newtonischen Grundsätzen abgehandelt.*[4] In this work, which Coleridge read, he proposed a hypothesis whereby an ordered cosmos was formed from chaos by the action of attractive and repulsive forces upon matter. The constitution of matter came to seem problematic. Were there discrete atoms, or was there a material continuum? In the second antinomy in his *Critik der reinen Vernunft,* Kant proved both the infinite divisibility of matter and its antithesis, the discrete monadic character of matter.[5] His *Metaphysische Anfangsgründe der Naturwissenschaft* (Riga, 1786) sought to give an a priori account of matter and motion. Kant argued that matter, defined as the movable in space, filled space by its moving forces of attraction and repulsion, which in themselves constituted matter. The boundaries of bodies were determined by the repulsive force, which was active only for a certain distance from its origin. Attractive force, however, "extends itself directly throughout the universe to infinity, from every part of the same to every other part."[6]

Coleridge admired Kant for this "first attempt," which he had had a hard time reading: "A really formidable sum of puzzle and perplexity, and repetitions of reperusals, with groundless apprehension: that my utmost efforts had failed to understand the writer's meaning, would Kant have spared me, had he but commenced the work with the plain avowal – that not the truth of Nature but the forms under which it may be *geometrically represented,* was its Object." Unfortunately for Coleridge, Kant was not uniquely concerned with geometric representation. Nor did he share Coleridge's distinction between body and matter. Finally, Kant, having derived matter as constituted by powers, proceeded to discuss matter as the subject of these powers. Coleridge believed that these were defects that he had avoided in his own system. "Let me not, however, fail to acknowledge, that a great Idea and worthy of Kant is contained in the construction of matter by two powers, the one [attraction] universal and the same in all, the other [repulsion] gradative and differential, and thus in each degree the ground of a specialty in matter."[7] Coleridge read this work more than once. His critical reading of it, however, came in 1819, after his reading of pertinent works by Schelling and Steffens, who were themselves indebted to Kant.

Schelling gave a brief and indeed cryptic statement of his theory of the construction of matter in his *Einleitung zu seinem Entwurf eines Systems der Naturphilosophie* (1799).[8] The first stage was the limitation of productivity by its opposite, as in the opposition of magnetic poles. Next came the tension between expansion and contraction, manifested in electricity. The third and final stage came with the transition of this polarity into a dynamic product or indifference, which revealed itself in chemical phenomena. As Coleridge wrote, probably in 1810, "Galvanism is the transition of Electricity into Chemismus [i.e., the power underlying and constituting chemical processes] or the co-adunation of Magnetism and Electricity."[9] Schelling concluded that magnetism, electricity, and chemical reaction were the categories of the original construction of nature, and thus furnished the general schemata for the construction of matter. This is at best elliptical, and barely intelligible.

Schelling, however, knew what he meant, and in 1800 published a fuller and clearer statement in the short-lived *Zeitschrift für spekulative Physik* that he and A. F. Marcus edited.[10] In this essay, he defined the construction of matter as the unique task of natural science. Given that magnetism, electricity, and chemical reaction were the general categories of physics, the problem was to show how these three functions alone enabled one to complete the construction of matter. He admitted that he had not shown this in his previous publications. He began his demonstration by postulating attractive and expansive forces in nature as an aspect of universal polarity. These forces could be represented as the opposite ends of a straight line. The generation of a straight line became, for Coleridge and for the *Naturphilosophen*, a powerful metaphor for productivity in nature. Schelling argued that in nature a line required three points for its definition, the end points or poles, + and −, and the midpoint or point of indifference, 0. This expression of the law of polarity, + ——— 0 ——— −, constituted the dimension of length. The corresponding physical force in nature was magnetism, with its poles in dynamic opposition. The magnetic needle or the bar magnet served as a symbol of the dimension of length and represented the first stage in the construction of matter. Schelling went on unconvincingly to cite experiments by Antonius Brugmans (1732–89) and others to show that magnetism acted only in the dimension of length. So much for the first stage, and the first dimension.[11]

The second stage, in which a second dimension, constituting breadth, was added to the first, was symbolized in nature by electricity. Electrically charged bodies were indeed electric over their whole surface, as Coulomb had shown. Static electricity thus illustrated the second moment in the genetic process that was the construction of matter.[12]

For matter to exist, a third stage or dimension was needed. Attraction and repulsion did not in themselves yield three dimensions; so Kant's construction in the *Metaphysische Anfangsgründe der Naturwissenschaft* was imperfect. The addition of a third dimension, constituting depth, came through the power of depth, the inward power of matter.[13] Coleridge, critical of Schelling for being too literal in his view of the third dimension, argued that depth was "an Idea, or as it were a spiritual intuition – not an image like length & Surface. In all these constructions Schelling pre-supposes the Space, and (what is worse) the Idea of *Depth* as ⟨if it were⟩ given by Space. But this polarization of Space into *the* space, this *substrat[ion]* of the idea, Depth, under the *image* of relative Lengths . . . is the very Knot to be unfolded."[14] Coleridge's point is that dimensions cannot be added to one another like bricks in the construction of a wall, but only, as Schelling generally insists, as part of a dynamic and genetic process; and the dimensions in this process correspond to constructive powers, rather than to merely static spatial dimensions.

Steffens, a disciple of Schelling, took the triplicity of this construction and converted it into what was primarily a quadruplicity, even while it held within itself the possibility of triad and pentad. Steffens, in short, in his *Grundzüge der philosophischen Naturwissenschaft* (Berlin, 1806), offered Coleridge a connection between Schelling's construction of matter and the pentad of powers in the compass of nature.[15] Steffens represented being and becoming by lines that he identified with the dimensions of length and breadth, respectively. Their intersection, identified with Schelling's third dimension of depth, was their indifference.

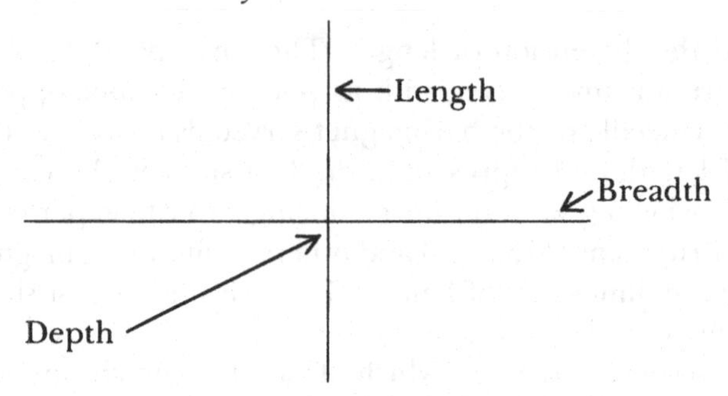

The identification of poles with compass bearings north, south, west, and east and their correlation with pairs of polar forces (north–south as magnetism, west–east as electricity) originated the symbol of the compass of nature. Coleridge, imbued with philosophical and theological trinitarianism and correspondingly critical of the *Naturphilosophen*, made Steffens's diagram the central symbol for a new pentadic logic with its inner triplicity (north–south, west–east, and their intersection) and its all-embracing unity. Triplicity in unity was the key:

> All things in Heaven & on Earth & beneath the Earth are but ~~as~~ one Triplicity revealing itself in an endless series of Triplicities, of which the common Formula is A + (−A) = B. – Deep was the aphorism of Heraclitus – that all was ⟨1, 2, 3 = 1⟩ ~~one~~ 1 & its opposite, and a third of both, & in which both were one. – that in *nature* its first manifestation, being that of *nascent* reality, was known by its *act* alone, = Magnetism = mere Length; its second ⟨a phænomenon⟩ πυροειδες τι [something firelike], the one disparting in duplicity = Length + Breadth = Surface = Electricity! – The interpenetration of the absolute opposites (which could not *be*, & yet be absolute *opposites*, if they were not the manifestations of *one*), & the perfecting synthetic Third, = Depth, = Gravitation = Galvanism = Chemical Combination – ![16]

The same symbols are used for the construction of matter and of the world. Coleridge, stressing triplicity, referred the creation of nature to an initiating act of the divine will rather than to an inner necessity; and he insisted that God the initiator and creator was a unity before his creation of polarity in nature. By September 1818, Coleridge, having in his *Biographia* been close to Schelling, had left him far behind, even while continuing to use his vocabulary in the construction of a dynamic scheme of nature.[17]

Schelling had erred in making nature absolute. This, from 1818 on, was Coleridge's constant and fundamental objection against his theories. And yet Coleridge, even when most critical of Schelling, was in sympathy with his enterprise.[18] A rational account of the construction of matter took one behind appearances to ideas. In the *Theory of*

Life and in notes and correspondence both before and after 1818, Coleridge used the metaphor of length, breadth, and depth.[19]

A decade later, he observed that it "is expedient, as soon as the evolution of Matter has reached to the Organic, to exchange the terms Materia or Hyle and Dynamics for Product and Productivity."[20] But the metaphors used in accounting for the construction of matter had their origin in the concepts of product and productivity, the keys to Schelling's cosmos and to Coleridge's. The metaphor of production, the point producing itself into a line as the first act of differentiation in the construction of nature, is taken from geometry. Euclid's second postulate in his *Elements* asks us to grant "that a terminated straight line may be produced to any length in a straight line."[21]

The indebtedness to geometry is not coincidental. Coleridge stressed it, as Schelling had done. In the *Biographia Literaria* we are told that "the word postulate is borrowed from the science of mathematics. . . . In geometry the primary construction is not demonstrated, but postulated. This first and most simple construction in space is the point in motion, or the line. . . . Geometry therefore supplies philosophy with the example of a primary intuition, from which every science that lays claim to evidence must take its commencement."[22] The phenomenal physical powers of magnetism, electricity, and galvanism served Coleridge as symbols of noumenal ideal powers, which were to be apprehended by reason. This, Coleridge wrote, was why "Plato and the Platonists" in their works enjoined geometry "as the ⟨first⟩ purification of the mind, the first step towards its emancipation from the despotism and disturbing forces of the senses."[23] The geometer and the philosopher of nature both began with ideas, so that geometric metaphors were especially apt, informing Coleridge's trinitarian logic.[24]

The role of triplicities, mediated by writers from Newton to Steffens, embodying geometric metaphor and illustrating dynamic philosophy, appears most strikingly in a notebook entry of the mid–1820s.[25] Coleridge began this note, entitled "Genesis and ascending Scale of physical Powers, abstractly contemplated," with

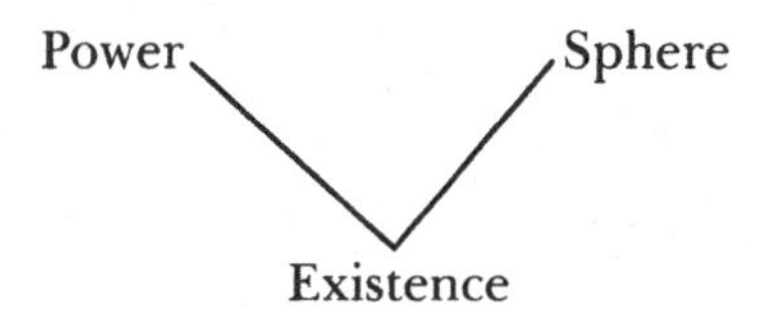

This was perhaps an assertion within the broad tradition of Newtonian dynamism that matter existed by virtue of its powers, and was limited by the extent of their sphere of action. Then came

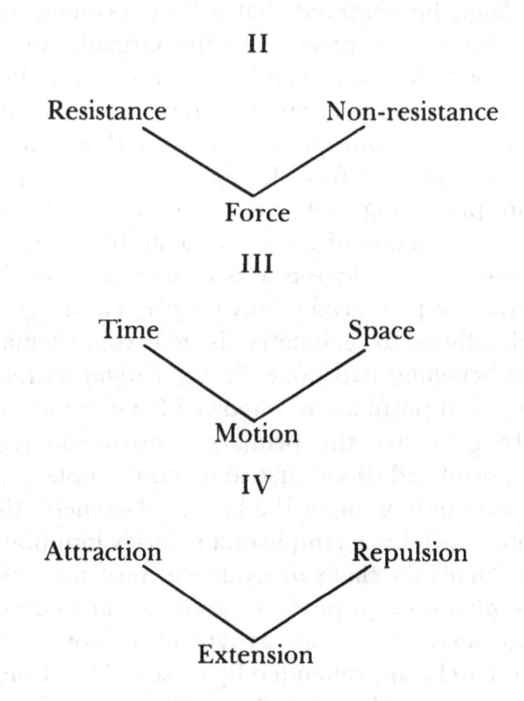

These triplicities appear to be related to Coleridge's reading of Kant's *Metaphysische anfangsgründe der Naturwissenschaft*. The next triplicity,

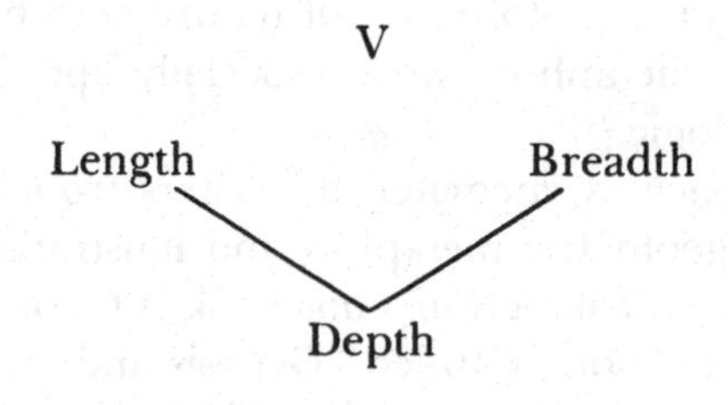

needs no further explanation, nor does

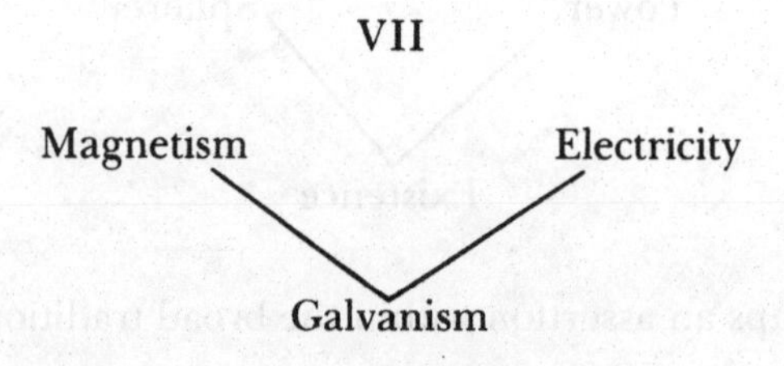

Schelling and Steffens have taken over from Newton and Kant. There were other triplicities, the last one being inserted as a post-script:

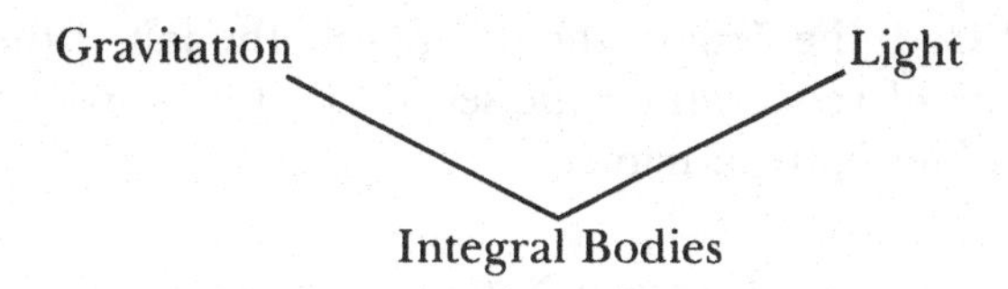

This, inspired by Moses and Steffens, would emerge as the key triplicity in the construction of the world of nature. The first two powers or members of each triplicity produced the third, "not as an Acid & Alkali *become* a neutral Salt, but as Parents produce *another* without ceasing to exist in their own persons."

In spite of the multiplicity of triads, length, breadth, and depth provided Coleridge with the primary symbols for his constructions. In his *Opus Maximum*, for example, he used geometric metaphors to express philosophical insights in mineralogy.[26] The scheme seemed universal in its applications. It needed, however, to be used with care, first in order to avoid the fundamental errors of the *Naturphilosophen*, and secondly to avoid simple confusion. A. F. Marcus, for example, had written an essay on inflammation in which he argued that the veins represented breadth and the arteries length. Coleridge was sternly critical of such meaningless confusion.[27] The scheme had to be understood precisely and applied consistently – no simple task! "Nothing but evil," Coleridge wrote, "can ensue from the doctrine of the 3 Dimensions, or even from the triple 3 – if it be not used chiefly as the basis of more special relations and oppositions. Ex. gr. Worse than vain would be the opposition of the arterio-muscular to the Nervous or to the veno-glandular powers, if at least equal attention were not given to the opposition in the Irritability itself between the muscles and Arteries, the firmamental and the fluid."[28]

Philosophical and Mosaic cosmogony

In 1820, Coleridge recorded his progress in the development and exposition of his "System of Philosophy and philosophic Science taught in a series of Conversations during the years 1817–1820." He had advanced in these years from cosmogony through geology to chemistry at the threshold of physiology and zoology.[29] In about 1815, as his notebooks and his marginalia to Steffens's works show, he had adumbrated his scientific program and had embarked on its early stages. He developed his cosmogony in its essentials in 1815–19. Almost a decade later, in May 1828, he placed on record a synopsis of his "Philosophy of Epochs and Methods," of which the first part discoursed of God, the Trinity, the divine will, the fall of the angels, and the scheme of redemption; the second part comprised metascience and the philosophy of nature.

> Part the Second
>
> Division 1. – Birth of Time and Nature by the Polarization of the Chaos. – Commences with the most perfect Contrary of the Absolute Act, that is at all predicable, or concerning which aught can be predicated. Division Second. – Polar Forces. Division Third. – Hints towards & hasty Sketch as in a Vision of the Forms of Nature as inorganic. – Division Fourth. Vegetable life. Division Fifth. Arguments for the possibility of the Mosaic Intervention of the Helioplanetic or Centri-peripherical Formation between the Manifestation of Vegetable, and the Birth of animal Life. – Anticipations of a New Science, viz. ~~Sc~~ philosophic Astrology, or Chemie Celestè[*sic*] – bearing ~~the~~ a similar Relation to the Newtonian Astronomy as Chemistry to Mechanics.

This fifth division is not present in Coleridge's earliest arrangements of sciences and powers. It emerged only after his essential rejection of Schelling's cosmology and his divergence from Steffens, being predicated not on *Naturphilosophie* but rather on trinitarian logic and biblical exegesis. The reference to "Chemie Celestè" was a deliberately transmuted echo of Laplace's celestial mechanics. Coleridge preferred sciences based on qualities to those based on quantities in attempting to construct his account of the cosmos. He went on to indicate the range of the "Remaining Divisions. – Animal Life from the Polyp to the primæval Man – Ends with the physiological, & the rational Grounds for the Assumption that Man is not in the state, in which the original Family must have been constituted and circumstanced: or a Fall of Man shown to be necessary Postulate of Science . . . This Second Part comprizes the prior half of the Exposition of the IDEA . . . that Life begins in detachment from Nature and ends in unition with God."[30] Coleridge's synopsis of 1828 gives a context to the cosmogony of 1815–19. This itself, as so often with Coleridge, embodies ideas at least partially anticipated by a "foreseen conclusion"[31] – a familiar situation mocking efforts at tidy linear chronology.

Cosmology for Coleridge embraced all the sciences, and its ideal form corresponded to his scheme of powers. If cosmology were to have scientific status for Coleridge, its theories and laws had to offer explanations whose terms were logically related through cause and effect. Tracing the causal chain backward brought one to origins, so that cosmology led one to cosmogony. Coleridge in his Bristol notebook defined cosmogony as "a System of the origination of all things including an explanation of their present state."[32] The *Naturphilosophen* subsequently gave him models of sufficient scope and hubris. In 1815–19 he adopted and modified these models to construct his own cosmogony.

By mid–1817, Coleridge had rejected the implications of Schelling's and Steffens's schemes, while remaining in sympathy with their aims. The grounds for his divergence from German philosophy were primarily theological and only secondarily logical. The first book of Genesis became the principal text organizing his philosophical cosmogony and distinguishing it from its German antecedents. The Gospel of Saint John and the Hermetic and Paracelsian doctrines embodied in the writings of Jacob Boehme were also important here, leading Coleridge to explore sound and color as a subsidiary polarity in his scheme.[33] But light and gravitation constituted the seminal polarity for him.

"In the beginning God created the heaven and the earth." "God the absolute Beginner of all things – not of their form, and relations alone, but of their very *Being*," Coleridge noted. "This Verse," he concluded, in 1818, "precludes the systems of Cosmotheism, Pantheism, and a primal Element – in short, every possible form of the eternity (i.e. unbeginningness) of the World or any part of the World."[34]

The second verse was more complex: "And the earth was without form, and void; and darkness was upon the face of the deep. And the spirit of God moved upon the face of the waters." The first part of this verse defined the primal chaos, Coleridge's dynamic prothesis that preceded polarization in the process of creation. Chaos was a "state of Indistinction," "an *absolute* Indistinction," which was an expression of pure potentiality. The Bible tells us that creation was God's act, and that chaos was its first stage. Coleridge saw this as logically necessary: "In order to *comprehend* and *explain* the *forms* of things, we must imagine a state *antecedent* to form . . . The requisite and only serviceable fiction, therefore, is the representation of CHAOS as one vast homogeneous drop!" And the second verse of Genesis was doubly wonderful to Coleridge because he saw it as embodying "the necessary form of Dialectic, or the evolution of Truth by means of logical Contradictions"; for chaos, being without form, was without surface or depth, and yet was described as having a surface – "the face of the waters" – and depth – "the deep." A prothesis embodying contradictions could serve as the ground for polarization and subsequent syntheses, in conformity with dynamic logic.[35]

Coleridge interpreted verses 3–5 as describing the initial act of polarization: "And God said, Let there be light: and there was light. And God saw the light, that it was good: and God divided the light from the darkness. And God called the light Day, and the darkness he called Night. And the evening and the morning were the first day." Here was the polarization of chaos into light and darkness, and with it the transition of nature from pure potentiality to actuality: "The actualizing principle has been infused."[36] Coleridge equated the Mosaic darkness with gravitation, so that the first differential act of the Logos could be presented as the creation of light and gravitation. "Now when the cosmogonist supposes, that . . . Chaos . . . being polarized (i.e. actuated by the Creative Will, the informing Word, and the disposing Spirit with two opposite correlative Forces) Light and Gravity became . . . , he . . . speaks of both *dynamically* – . . . Gravity as the Vis Massifica, Natura *se inhibens*, et uniens – the real or objective Pole of Nature: Light as the *distinctive* Power; Natura evolvens ab intra, evocans ab extra, et [?omne modo] *se exhibens*: the Ideal or Subjective Pole of Nature."[37] The distinction is essentially logical for Coleridge, but it represents the commencement of the law of polarity, and thus makes possible the development of philosophical cosmogony. The reference to subjective and objective poles reminds us of his indebtedness to Schelling's vocabulary. But Steffens, for all his errors, was the more immediate inspiration. Coleridge regarded both light and gravitation as bipolar, and correlated them with the polar lines west–east and north–south, corresponding to electricity and magnetism and also to breadth and length in the compass of nature.[38]

Steffens in his *Grundzüge* constructed his compass of nature, claiming that it enabled one to orientate oneself in apparent chaos and that it arose from the relative opposition between becoming and being, "that is, between Gravitation and Light." He went on to assert the correlations that Coleridge adopted: "North–South – being – under the powers of gravity – magnetism – line. West–east – becoming – under the power of Light – electricity – surface."[39] Other *Naturphilosophen* built their schemes from the same concepts. For example, Adam Karl August Eschenmayer (1768–1852), professor of medicine and philosophy and then of practical philosophy in the University of Tübingen, used Schelling's triplicities based on magnetism, electricity, and galvanism in constructing an account of nature, with the polarity of light and gravitation as the basis of his cosmology and theoretical astronomy. Coleridge read Eschenmayer's *Psychologie* with critical care, and it influenced his own theory of the heavens.[40]

The first verses of Genesis, Coleridge's dynamic logic, the concept of the Trinity, Schelling's view of nature, and Steffens's compass of powers all contributed to Coleridge's cosmogony, in which the polarity of light and gravitation was fundamental. Light and gravitation were conceived as noumenal powers, whose polarity provided the model for all logically subsequent polarities.

A principal task of Coleridge's philosophy was the evolution of "all the powers that now are . . . each in its own branch in the tree of philosophic genealogy, the great pedigree of the world," predicated on the prior polarization of chaos into light and gravitation, the origin of all the forms and powers of the universe.[41]

Coleridge's wrestling with German philosophers and his pursuit of a philosophical exegesis of Genesis are apparent in the manuscript of his *Opus Maximum.* The first of the three volumes of this work in Victoria College begins, typically, with a supposed digression on cosmology and cosmogony. This "digression" then turns itself into a discussion of gravity, light, the Logos, and powers, which concludes "by a distinct recapitulation and specification of the powers hitherto evolved in the order of their birth and epiphany." "Mere potentiality . . . i.e. chaos philosophicum" was first actualized, yielding the indistinction or darkness that was upon the face of the deep. Then came the creation of light, "Verbum lucificum . . . the Multeity actualized." The polar conjunction of light and darkness followed, and thence the development of the material creation. Polarity appeared in three principal forms. There were first the poles "realized . . . and manifested in the Creature – Vis centrifuga ♓ Vis centripetalis," corresponding to ordinant powers, and belonging to the entire mass of bodies. Next came the "actualized Indistinction ♓ Multeity. Offspring in the Creature or the realized Poles, Attraction ad extra . . . or Astringency ♓ separative self-projection or Volatility," corresponding to substantive powers, and belonging to the components of the mass. Finally there were modifying powers, belonging to "the differences superinduced on the . . . components." Coleridge represented this third polarity by "Particularization, Contraction, as ♓ Omneity, Dilation."[42]

He was here using familiar words with unfamiliar precision that transmuted them to new coinage. He had his reasons for this, and had good precedent for his action: "In the Preface to Paracelsus' Works extract the eloquent defence of technical new words, & old words used in a new sense." And again: "A philosopher's ordinary language and admissions, in general conversation or writings *ad populum,* are as his watch compared with his astronomical timepiece."[43]

Having categorized polarity according to ordinant, substantive, and modifying powers, he set about showing how the "play and changes of these form the physical contents of the history of the Cosmogony." For example, he argued, suppose that the equilibrium between volatility and astringency were modified by contraction, and that the centripetal force were added. Then, according to his scheme, the result would be cohesion. Dilation could be similarly generated. When volatility had mastered the indifference between contraction and dilation, the phenomenon generated in correspondence to the resultant power would be – water.[44] Coleridge's demonstrations here were elliptical and perhaps wonderfully improbable. But they were not arbitrary. They had their rules and their logic; the terms in their several meanings were precisely defined, and were represented by a self-consistent set of symbols[45] for whose use Coleridge adduced philosophical, theological, and scientific justification. The logical construction of powers is basic to Coleridge's dynamic cosmogony, rendering it at once rational and genetic. It is also the region in which he is closest to Swedenborg and furthest from ourselves. His classification of the sciences and of natural phenomena in terms of powers is, however, readily intelligible, productive of insight, and suggestive of fruitful areas for investigation.

The book of nature and the Book of Genesis

A logical derivation of powers was one of the goals of the *Naturphilosophen.* Steffens used dynamic philosophy together with geological and chemical lore to construct a dynamic geogony. Coleridge criticized the philosophy; he rejected specific conclusions, comparing Steffens's account unfavorably with Genesis: "How much more simple is the brief Mosaic account," and again, "Moses is a better Teacher." Yet he remained sympathetic toward Steffens's "dynamic and vital Schematism of the terrestrial and planetary phaenomena." Moses might prove the better teacher. But Steffens provided the vocabulary and a model for a philosophical exegesis of Genesis.[46]

This exegesis succeeded triumphantly for Coleridge, the first verses of Genesis emerging as ever more pregnant with truth and rich significance.

> Thus from Magnetism, Electricity, and their productive Synthesis (= constructive Galvanism or ~~Chem~~ Alchemy) God εκοσμησε το Χαος [rendered the Chaos into Cosmos], and the following Verses to the imbreathing of the *Soul* into Man relates the process as far as the appearances or human Perspective is concerned –. Hence the successive products are related in the order in which they would have been noticed by

a human spectator, and in the order in which they took place only as far as the latter coincides with the former – in that part only scientific, in which pure Acts are related that could not have been made sensuous – but in this part profoundly scientific. – Hence too the Sun & Stars are placed in the same Firmament (literally & rightly, Expansion) with the vapors, or waters above the firmament, which doubtless must have then borne a vastly greater proportion to the Sea than at present – Hence the Insect Tribes which probably were synchronous with the vegetable creation are post-dated to that of the Animals, and the Zoophytes with the lowest initia of animal & vegetable Life, probably the continuing product of the ~~firs~~ second, third, and fourth Days or Periods are not mentioned at all. – So that falsehood is nowhere to be found in this sublime Hymn.[47]

Coleridge's philosophy and his ventures into science were contingent on his religion; some congruence between his own cosmogony and the one he read in and into Genesis is accordingly to be expected. Yet Coleridge sought secure and detailed scientific foundations for his system, and the strength of a framework at once logical and rational. He felt that he had succeeded, that his cosmology rested on his doctrine of powers, including the difference of "Gravitation both from the centripetal Power & from simple Attraction" – and that he had arrived at this conclusion "by necessary evolution from the First Principle of my Philosophy before I was aware of it's exact coincidence with the Mosaic Cosmogony."[48] As Coleridge extended his learning, he found ever more grounds for belief in the scientific accuracy of the story of creation in Genesis.[49] In a marginal note to a passage in which J. G. Eichhorn wrote of the majesty and truth of the Mosaic account, Coleridge wrote decisively: "So *I* think, who take Gen. I literally and geologically: and so *did* I think, when I interpreted the chapter as a Morning Hymn."[50]

It was easy to move from the discovery that Mosaic cosmogony could be reconciled with the latest and most recondite discoveries and theories of science to the view that the first chapter of Genesis was a fertile source of scientific insight. The symbiosis between the laborious reconstructions of the moderns and the inspired account in the Bible was complete for Coleridge by 1826:

I am most thoroughly aware that the Bible was given for other & higher purposes than to make us Naturalists: that God never intended to supersede Industry by Inspiration, or the efforts of the rational Understanding by the Revealed Word. Yet I am not ashamed to confess that I am strongly disposed to regard the first Chapter of Genesis, as in some sort an exception: as providing certain Truths a priori, as the supposita et postulata, vel potius jussa et præfinita of Physiogony, which could have been ascertained . . . by no lower authority.

One of these basic truths, informing Coleridge's philosophy of nature no less than biblical cosmogony, was that "the World proceeded from the less to the more perfect, from indistinction to Order thro' [a] numbered ascending series of distinct ~~and~~ evolutions . . . – Beneficial Results of this exemp. in Kant's Himmel-system, as developed by Le Sage & La Place."[51] Modern cosmology, like that of Moses, evolved from chaos to order.

Coleridge strove to bring together science, revelation, and a philosophical account organized through a system of symbols.[52] His science was always one of powers, not of things – "for I have nothing to do with Elements, but follow Moses." Establishing the concordance of science and Scripture thus became an integral part of Coleridge's enterprise, a necessary prelude to the *Opus Maximum*, but one not without difficulties. Genesis 2:5, for example, states that God made "every plant of the field before it was in the earth," whereas Genesis 1:11 states that "God said, Let the earth bring forth grass." Coleridge reconciled the two statements by assuming that Genesis 2, "probably the elder and ruder document speaks of the *Germs* as in the air" – a labored rescue operation.[53]

This reconciling exegesis was truly a labor for Coleridge, and one in which he persevered. In 1829, when the concentrated study of natural science was behind him, he summed up endeavor spread over more than a decade: "It is scarcely possible, I say, [that] any . . . man can more earnestly desire than I do, to find the first ten Chapters of the Book of Genesis true, historically or symbolically / either as narrating past Facts or conveying perpetual truths. Accordingly, no man has labored with more zeal and perseverance to find or to recognize truth, scientific, religious and anthropological in these Chapters – and, I humbly thank the Father of Lights! not in vain. / But unmixed with error, . . . – to this state of Assurance I have not yet attained."[54]

The details might be tentatively held and open to revision. The program of interpretation remained constant throughout the 1820s. "It seems to me clear," Coleridge wrote in 1829, "that Ch. I to v. 4. C. II. is a scheme of Geogony, containing facts and truths of Science adapted to the language of Appearances and the popular notions grounded on Appearances – It is throughout liberal – and gives the physical Creation."[55] By 1829 scientific arguments had extended the age of the earth well beyond the six-thousand-odd years of biblical fundamentalists. Coleridge, convinced of the unity of scientific and religious truth, was undismayed. "The Mosaic Cosmogony," he observed, "requires nothing more then the recency of the conclusion of its 6th great Epoch or Day. You may push back the ⟨posternal⟩ Vesper of its first Day . . . as many myriads or millions of Years, as you like – or as the facts of Oryctognosie may suggest."[56]

The world before animal life

So far we have advanced with Coleridge no further than the end of the first day of creation. This was the archetypal act of creation, the model of "the eternal act of creation in the infinite I AM" itself the model of the activity of Coleridge's primary imagination.[57] Divine will has acted, the initial polarization from chaos has been produced, and the creation of light and darkness has served to initiate the evolution of a succession of polar powers.

The next five verses of Genesis (vv. 6–10) present the creation of "a firmament in the midst of the waters," which God called Heaven, and the creation of the earth and the seas. This is followed by the creation of vegetable life in all its forms (vv. 11–12). Life on earth emerged after the differentiation of solidity and fluidity. Here for Coleridge was beautiful imagery and sound science:

> Water the noblest material Image as out of which all the Forms of Life issue and into which they return . . .
>
> Doubtless, the organic and the inorganic are necessary relative distinctions, but as phænomena only. – the lightest changes even in the mere relations of Space are followed by pulses of Light, Warmth & electricity – so compleatly interdependent, each end and means at once, is the animated Nature!"[58]

With the creation of heaven and earth, of air, land, and sea, we have the necessary prelude to the creation of life. We also have a world that can be described in the language of geology, and in the language of chemistry, whose substances were for Coleridge exponents of powers. Now Coleridge sought to achieve an integrated understanding, in which geological, chemical, and cosmological relations were complete and clear.[59] The sequence of powers in their evolution, the hierarchy of sciences erected on that foundation, and the Mosaic account would then come together, enriching our understanding of each.

Coleridge attempted this task, with a fair degree of success, in notebooks from 1818 into the 1820s, and most coherently in the first volume of his *Opus Maximum*. Man is the culmination of creation; the organic powers are the last and highest in Coleridge's sequence; and the organic sciences, notably physiology, are at the top of his hierarchy. But there are complications. First, the Bible places the creation of vegetable life before the creation of the sun and planets. Life and the life sciences thus occur at two separate states in the development of the cosmos, which is itself organic. Coleridge recognized this, and yet his schemes of the sciences place the vital and organic realms entirely after the cosmical and geological ones.[60]

It is tempting to seek to dissolve rather than resolve the inconsistency by appealing to his distinction between historical and logical treatments of the evolution of powers. But this merely compounds the problems. Coleridge's tables of powers and sciences are not all identical, nor are their changes linear with time. More fruitful in achieving an understanding of these inconsistencies is the recognition of the resonance between the levels in his hierarchy of powers. The vital realm has its chemical aspect; the chemical realm operates with higher powers than geology, while yet working its effects in geogony; and galvanism provides a metaphor for chemical and vital activities. The interdependence of the different realms made it seem reasonable to Coleridge that the scientifically educated philosopher should aspire "to see a World in a Grain of Sand."

The creation of vegetable life was the last act of the third day. The fourth day saw the creation of sun and moon, stars and planets (vv. 14–19). The creation of animal life occupied the fifth and sixth days, whose climax was the creation of man. The greatest triumph of Newtonian science had been its precise quantitative account, in the writings of Laplace, of the mechanism of the heavens. Coleridge found Schelling's explanation of the world as organism much more congenial, and denied the supremacy of mathematical physics. He wanted to explain quantity as derived from quality. The *Naturphilosophen* had pointed the way toward doing this. The Bible symbolized and reinforced his scheme, first in deriving powers from chaos, and then in showing in cosmology how organism – vegetable life – preceded mechanism in the planetary system.[61]

Coleridge was no mathematician:

> With what bitter regret, and in the conscience of such glorious opportunities . . . [to learn mathematics] all *neglected* with still greater *remorse*! O be assured, my dear Sons! that Pythagoras, Plato, Speusippus, had abundant reason for excluding from all philosophy and theology not merely practical those who were ignorant of Mathematics . . . I cannot say – for I know the contrary . . . – that [true knowledge] *cannot* be *acquired* without the *technical* knowledge of Geometry and Algebra – but never can it without them be adequately *communicated* to others – and o! with what toil must the essential knowledge be *anguished-out* without the assistance of the technical![62]

The books of mathematical astronomers were thus virtually closed to him, except in their discursive parts.

Cosmology and astronomy

Coleridge accordingly had problems in understanding astronomy, with which he wrestled optimistically, seeking to bring this refractory science into his scheme of knowledge. A notebook entry of 1830, related to his collaboration with J. H. Green, exhibits his optimism, his encyclopedic intellectual ambitions, and the all-but-insuperable difficulties in their way: "Mem. If I can but get satisfactorily thro' the inorganic, the Vegetive, and the astronomical – and we are fairly landed on the coast of Animal Life, I see plainly & with the delight of a confident anticipation, that we shall thenceforward advance pari passu, propelled and propelling."[63]

Coleridge's interest in astronomy was certainly long-standing. Before going to Germany he had encountered Newton's works and also the crumbs of cosmology in Erasmus Darwin's writings. His gleanings from Darwin included a hypothetical explanation of the origin of the moon, thrown from the earth "by the explosion of water or the generation of other vapours of greater power." Some two decades later, reading G. H. von Schubert's *Ansichten von der Nachseite der Naturwissenschaft,* he developed this idea in a marginal note, suggesting "that our Moon was a detachment from the Earth, a sort of After-birth projected during or towards the close of the Orgasm of Solidification." Other marginal notes in this work make it clear that Coleridge was also aware of the subject matter of William Herschel's papers on double stars, nebulae, and comets, all published in the *Philosophical Transactions of the Royal Society of London.*[64]

Coleridge depended for the most part on reviews and textbooks, for example, Roger Long's *Astronomy* (5 bks. in 2 vols., Cambridge, 1762–64), and even here encountered difficulties: "Mem. To inform myself what is meant by the Magnetic Period, Equatic &c."[65] He read the reviews – for instance, the "Recent history of astronomy," *Quarterly Review* (*38* [1828], 1–15), and still found problems. The reviewer, discussing J. F. W. Herschel and J. South's work on double and triple stars, mentioned that there were probably stars that revolved around each other. Coleridge wondered what that meant, guessing – correctly – that they revolved around their common center of gravity. But "at what distances?"[66] His interest in astronomy, far exceeding his competence, was that of an outsider. When William Rowan Hamilton presented him with an essay on mathematical optics and astronomy, it was in token of respect and regard, not in hopes of being read and understood.[67]

There are notebook entries and marginalia indicating skill in precise astronomical observation – for example, of the form of the Milky Way – and an interest in striking phenomena, such as a total eclipse of the moon.[68] Coleridge's interest in astronomy was, however, largely a philosophical one. The *Naturphilosophen*, especially Steffens, Schelling, Eschenmayer, and G. H. Schubert, pointed toward the possibility of a new science, "philosophic Astrology."[69] Here, however, as always, Coleridge required an accurate foundation in nature before the construction of philosophical science. Comets furnished an important part of that foundation. Their manifestations were striking, they were readily and regularly observed, and the resulting literature was accessible and controversial. Newton had determined their orbits in the *Principia*, and both before and after this there had been numerous theories about comets. Coleridge read about them sporadically. William Herschel published papers on comets in 1787, 1789, 1807, 1808, and 1812 – there had been "a particularly splendid and prolonged" comet in 1811–12. Coleridge annotated "A letter on comets, addressed to Mr. Bode, Astronomer Royal at Berlin," in 1809, and in 1810 read and made careful notes on the account of comets in J. H. Schröter's *Beiträge zu den neuesten astronomischen Entdeckungen*, ed. J. E. Bode (3 vols., Berlin, 1788; Göttingen, 1798, 1800).[70] He probably read the chapter on comets in Richard Saumarez's *Principles of Physiological and Physical Science* (London, 1812), which attacked the foundations of Newtonian philosophy.[71]

Then, in 1818 and 1819, Coleridge essayed a thoroughgoing critique of Newtonian explanations of the nature and behavior of comets. The date fits in well enough with the development of his system and plan of studies. He surely read Schröter's observations on the 1811 comet, published in *Blackwood's Magazine* (*3* [1818], 338). Coleridge responded, embarking on a series of informed queries. If comets were made of self-luminous matter, why did they increase in visible splendor as they approached the sun? How could a comet whose nucleus was as great as Jupiter fail to disturb one of Jupiter's satellites when passing close by the planet? If comets' tails were as tenuous as astronomers believed, how could they reflect light? These and other difficulties opposing themselves to Newton's theory were addressed to the editor of *Blackwood's Magazine*, probably in the fall of 1819. The letter, which was not published in *Blackwood's*, ended coyly: "I cherish, I must confess, a *pet* system, a bye blow of my own Philosophizing; but it is so unlike to all the opinions and modes of reasoning grounded on the atomic, Corpuscular and mechanic Philosophy, which is alone tolerated at the present day, and which since the time of Newton has been universally taken as synonimous with

Philosophy itself – that I must content myself with caressing the heretical Brat in private – under the name of the Zoödynamic Method – or the Doctrine of *Life*." In a marginal note to Thomas Stanley's *History of Philosophy*, organism became explicit: "Comets the products of the out-breathing of the System, and the organ of respiration correlative to inspiration?"[72]

Proponents of the mechanical philosophy had sought to explain the system of the heavens as a consequence of the properties of matter. Buffon in his *Histoire naturelle* explained the origin of the planets by postulating the collision of a comet with the sun. The matter thus torn from the sun's surface was then supposed to have condensed, forming the planets. This explained nothing for Coleridge, who seems to have misunderstood Buffon, believing him to have postulated without explanation the projection of planets from the sun. What in the nature of the sun could have caused this sudden exertion of repulsive force?[73] "If we suppose with Buffon, Schelling & others, that the planets &c were exploded from the sun, we have then to ask whence came the sun?" Mechanical philosophy proposed effects as causes.[74] It also claimed as imaginatively stupendous what was ultimately sterile, the size – not grandeur – of the universe, and its regularity. Kant in his *Allgemeine Naturgeschichte* had been impressed by both. Not so Coleridge, who found no "super-superlative sublimity" in a mechanical universe. The effect of contemplating Newtonian astronomy, a system of balls spinning in space around other balls, "is not only depressive from its monotony but revolting from its want of analogy to . . . all our other experiences of . . . Nature."[75]

His favored alternative was a developing world in which organism was the dominant metaphor. As early as 1810 he was asking hopefully "whether Reason does not command us to judge of these astronomical & geophysical necessities by the contrivances of the organized world, & not vice versâ." A notebook entry of around 1823 stated uncompromisingly that "Tellus is an organic Part of a System, the Solar, which itself is probably but a part." Coleridge became increasingly confident that the earth had undergone a process of organic growth; "nor can any reason be adduced why that which holds good of each Body of the Solar System should not be true of the System itself."[76]

The *Naturphilosophen* were here, as often elsewhere, at least on the right lines. Coleridge's revulsion against mechanistic cosmogony was almost a paraphrase of Gotthilf Schubert's contention that explanations based on atoms and attraction "directly contradict all analogy, indeed all true nature." And Coleridge's marginal note about the emergence of the moon during an orgasm of solidification, together with Schubert's earlier assertion in the same volume, *Ansichten von der Nachseite der Naturwissenschaft*, that everything went from featureless fluidity to an ever-more-solid state, must be seen in the context of physiological theories correlating the development of life with the emergence of solid organization from fluidity.[77] Eschenmayer's criticism of Kant for ascribing too little to the role of organism, and too much to mechanism and geometry, surely also met with Coleridge's approval.[78] For all his disagreements with Steffens, he acknowledged his sympathy with Steffens's view that the relation of the earth to the sun was a living one, and that the whole earth was held in an "organic compulsion" by living "Organismus."[79]

Newton's law of gravitation did yield a good description of planetary orbits. But it dealt only with universal gravitation, and not with specific gravity; it ignored the inner nature of bodies, symbolized by Coleridge as the power of depth or chemical power. Surely, he argued, Newton's gravity was a manifestation or product of the inner nature and power of bodies. Indeed, "so does the Chemique constitute Gravity as particular Gravity." If Newton had ignored "specific gravity," others, unnamed by Coleridge but presumably including Steffens, had built their geogony using it to the exclusion of universal gravity. But Coleridge argued that it was wrong so to divorce cosmogony from geogony. What was needed was a new unified science. Laplace's *Mécanique céleste* would be superseded.[80] Steffens had at least an inkling of this, invoking polarity and organism. And, here as so often, Giordano Bruno seemed to Coleridge to have been gifted with prophetic insight, regarding the stars as suns surrounded by their planets, and seeking "to deduce this a priori from centro-peripheric Process, or primary Law of Matter; which he elsewhere calls the law of Polarity."[81]

Coleridge knew of no one except Bruno before the *Naturphilosophen* who had sought to deduce the system of the heavens from the law of polarity. But he knew that several astronomers and philosophers – including Kant, the Swiss natural philosopher G.-L Le Sage (1724–1803), and Laplace – had attempted to construct evolutionary cosmologies. Kant in his *Allgemeine Naturgeschichte* assumed an initial even distribution of individual atoms throughout space and invoked Newtonian attractive forces to produce condensation of the atoms into stars, suns, and planets. Planetary systems thus formed rotated within disc-shaped galaxies that were themselves distributed like super-planetary systems within the super-galaxy of the universe. The history of the universe was for Kant one of regular motion and order arising from rest and chaos in accordance with god-given laws. Laplace's theory, presented in his *Exposition du système du monde* (2 vols., Paris, 1796), is also evolutionary, but is limited to an account of the solar system, postulating that the sun was originally a vast nebula rotating slowly and contracting as it cooled by radiation. As it contracted, it rotated faster, flinging off successive rings of matter in the plane of its equator. Each ring condensed into planetary bodies. Laplace's theory was different from Kant's, and probably entirely independent of it. But Coleridge, admiring Kant and noting the priority in time of his evolutionary cosmology over Laplace's, saw the latter as merely a development in detail of the former – a fair indication that he had never read Laplace. G.-L. Le Sage was a poorer mathematician and astronomer than Laplace, a poorer philosopher than Kant, and more of a mechanist than either of them; he invoked imponderable fluids as the cause of gravitation – an ultimate solecism in Coleridge's view, which put him only by courtesy with Kant and Laplace as an evolutionary cosmologist, while limiting the only tenable parts of his cosmology to those supposedly borrowed from Kant. When Schelling praised Le Sage's mechanical physics, Coleridge furiously attributed this compliment to the Frenchman to "base envy & jealousy" of Kant.[82]

Kant and Laplace had built their evolutionary cosmologies on the foundation of Newton's physics. Coleridge, however, saw the laws of the planetary system as Kepler's discoveries, believing that Kepler was a far greater genius than Newton. This was a viewpoint widely shared by the *Naturphilosophen* in Germany, who admired Kepler's dynamism, his search for unity and harmony, his recognition of the active role of mind in natural science, and even his efforts to deduce a priori the ideal geometric structure of the solar system.[83] Steffens presented a derivation of Kepler's laws that Coleridge worked through and criticized in 1819. Today, astronomers give Kepler's laws as follows:

1. Each planet travels in an ellipse, one focus of which is occupied by the sun.
2. A line drawn from the sun to a planet describes equal areas in equal times.
3. The squares of the periods of rotation of the planets about the sun are proportional to the cubes of their mean distances from the sun.

Steffens reversed this order, so that his first law was Kepler's third. His derivation was based, obscurely enough, on identity and polarity, and concluded: "Whereas the first Keplerian law expresses the identity of the planet with itself, and the second the relative difference, the third expresses the synthesis of identity and difference, and thus these laws express the entire type of regularity in the movements and being of the heavenly bodies as relative totalities, and in so doing completely express the type of being of the particular in the general." Coleridge had difficulties with Steffens's account. "But how can this be on the Identität-Lehre? . . . – This, however, is *their* concern. It may be illegitimately deduced from their system, and yet be Truth and a rightful consequence in ours. To work, therefore, to work! and first to state the 3 Laws of Kepler . . . " He succeeded in making out Steffens's version of the laws; their dynamic interpretation appealed to Coleridge, yet Steffens's derivation of them from the absoluteness of finite nature was repugnant to him.[84]

Steffens might therefore seem an unsuitable source for Coleridge in his attempts to reconcile Genesis with astronomy. But Steffens's approach to the history of the earth as essentially organic, and to the development of the planetary system as coincident and interdependent with the development of the earth, was compatible with Coleridge's philosophy and with Genesis 1:10–17. So he persevered with Steffens while considering the theories of other cosmologists. In a lengthy notebook entry of 1819–20, Coleridge presented his most nearly coherent cosmological enquiry, conducted in the context of a critical reading of Steffens's *Geognostisch-geologische Aufsätze* . . . (1810), and C. A. Eschenmayer's *Psychologie* . . . (1817).

Steffens, looking at the fossil remains of equatorial plants in northern anthracite mountains, argued that the growth of these mountains involved vegetative processes, which in turn implied a former spatial relation between the sun and the earth different from the one that now exists. Perhaps the angle of the ecliptic to the celestial equator had changed, with a consequent displacement of the terrestrial equator. Steffens noted that this interpretation, while incompatible with Laplace's perturbation theory, was compatible with geognostic evidence. He was thus led to the questions that were central to his theory: "What if the history of the formation of the earth coincided with the history of the planetary system? What if the same formative force that separated individual life on earth, and each life its own inner centre of being, was also the separative force for the planets?" Wasn't it possible, he asked, that what manifested itself in the "narrower circle of life" might also occur in the larger sphere, "so that the planets were reciprocally influenced more by polarity and outer opposition than by a general and central relationship?" Newtonian astronomy was irrelevant here, because it was not applicable to the early history of the world. Steffens went so far as to suggest the relation between the planets was then dynamic and qualitative rather than quantitative. Different relations of the sun and planets suggested that the years and seasons might formerly have been of very different duration from our present ones; the earth would have moved more slowly, and the seasons would have been longer, leading to the formation of huge alternating strata in the earth.[85]

But, as Coleridge pointed out, these strata were irregular, and not explained by any astronomical periodicity. The succession of mineral, vegetable, planetary, and animal creation in Genesis 1, corresponding to a succession of powers, provided him with an alternative explanation of these strata, without requiring regularity.[86] For all its attractions, Steffens's account was inadequate. Coleridge was still left with the problem of explaining the existing relation of sun and planets as later in time than the creation of the earth. It was now that he turned to Eschenmayer, student of animal magnetism, and one-time follower of Schelling. His *Psychologie* discussed cosmology in terms that had a partial congruence with Coleridge's philosophy.[87]

Eschenmayer hypothesized that no star was without its attendant planets, and that every solar system had a central focus that governed the whole, determining the relative distances, paths, and speeds of the other bodies. This established the idea of a *Naturzentrum* as the ordering principle for physical bodies, and even as causality in nature. Eschenmayer claimed that it could be demonstrated that there need be no body at the *Naturzentrum* for it to function. He proposed no mere equilibrium between bodies, but a center containing the rules for each individual relation in the system and serving as the originating center of action within the system. Coleridge was scornful of the way in which Eschenmayer derived the sun from the *Naturzentrum* while seemingly denying it body. But there was much in the system that he liked. To begin with, Eschenmayer identified his concept of a generative focus with the objectification of the free causality in will. Coleridge's cosmos was initiated by an act of will. Then, Eschenmayer claimed that the *Naturzentrum* contained the triplicity of the fundamental cosmic phenomena, light, heat, gravity. The concept of triplicity, of trinity, became increasingly important and ultimately dominant in Coleridge's thought. Eschenmayer's approach enabled him to get at the three fundamental qualities of nature without using the atheistic principle of identity. Also attractive was his discussion of the mutual attractions of the sun and planets, which were the result not so much of weight or gravity as of a chemical power.[88]

Coleridge's limitations in approaching the sciences are nowhere more evident than in astronomy. It was not possible to anguish out the necessary knowledge in ignorance of mathematics. He was quite unable to understand the significance of mathematical physics in general and of Newtonian astronomy in particular. As for the philosophical cosmologists he read, they either, like Steffens, began from premises that were unacceptable to him, or, like G. H. Schubert, whose *Allgemeine Naturgeschichte* (Erlangen, 1826), he annotated, were confused in their arguments: "But throughout Schubert too much confounds the powers and the eternal laws, that are the *conditions* of the actualization of the Powers."[89] Coleridge's inquiry into astronomy was foredoomed, like that of the mechanical philosophers before Newton, to a series of program statements, disconnected insights, and explanations in principle that lacked predictive power. It was not until he came to what in his day were truly sciences of quality – to chemistry, for example, and to physiology – that method, knowledge, and system united fruitfully. In physics, once beyond the philosophical generalities that he handled so surely, his critical inquiries were inevitably frustrated.

General physics

Coleridge, advancing philosophical grounds for developing the sciences through qualitative rather than quantitative concepts of power, was uninterested in the two principal versions of Newtonianism that had their origin in the thirty-first query to the *Opticks*. The first of these versions interpreted physical changes through the concept of quantifiable forces acting according to universal and verifiable laws. The alternative version embodied different kinds of activity in distinct imponderable but material fluids. Neither version met Coleridge's philosophical requirements.

The science of electricity had become enormously fashionable and intellectually exciting following Volta's discovery of the pile, the first man-made source of current electricity.[90] But nowhere in Coleridge's notebooks is Volta's name mentioned, for Volta wrote about electrical fluids. Galvanic and static electricity for Coleridge were embodiments of polar power symbolizing laws, just as they were for Schelling in his construction of matter. This view, combined with his ignorance of mathematics, led Coleridge to ignore the dramatic advances in electrical science made by Ampère and Coulomb in the early 1820s – precisely when he was most actively formulating his own schemes of the sciences. Thermal activity in nature had likewise been embodied in a fluid – caloric – by Lavoisier in his reformation of chemistry.

Subsequent studies in thermal conduction and radiation had taken advanced mathematical forms, whereas calorimetry, the measurement of quantities of heat, was firmly based on the fluid model.[91] It seemed to Coleridge that there might be useful fictions here, but no possibility of philosophical insight. He therefore neglected most contemporary advances in the theory of heat. In chemical reactions, however, heat was important for him. His criticism of the hypothesis of caloric may serve here as a prelude to chemistry.

Heat

In 1819, while working his way through William Thomas Brande's newly published *Manual of Chemistry*, Coleridge remarked on Brande's account of the production of oxygen gas by heating "the Black Oxyd of Manganese" to redness. He had also heard Humphry Davy describe the same process in his early lectures at the Royal Institution. Once again he was struck by the generality of thermal phenomena in chemistry, and wondered about incorporating them in his scheme of powers. "It is more and [?more] evident to me," he wrote, "that little can be done to any purpose in philosophic *Chemistry*, till the nature of Heat and Fire in their various manifestations be satisfactorily enucleated . . . At present, I should say conjecturally, that ignition implied the maximum of *antagonism*, the wrestling war-embrace of the Contractive and Dilative – and that melting = the superiority of the latter, and Gas that of the former."[92] This concern with the role of heat in chemistry was an ancient one, going back to Paracelsian and alchemical traditions. Its connection with the general question of changes of state and the powers of matter had recently come under investigation in two very different schools, which in Coleridge's day had their respective centers in France and Germany.

The German school, based on *Naturphilosophie*, considered heat in the general context of polar powers. The approach in France was quite different. Scientists there, for example, Barruel and other teachers at the Ecole Polytechnique, divided *physique*, the study of matter, into two principal branches, *physique générale* and *physique particulière*. The former dealt with behavior common to all matter, including mechanics, capillary action, and Newtonian astronomy. The latter dealt with behavior that varied from one kind of matter to another, and came increasingly to be identified with chemistry. But because in this scheme chemistry was a branch of *physique*, efforts were made, by C. L. Berthollet and Laplace, among others, to extend *physique générale* into the molecular realm, and to relate the general properties of matter to chemical properties.[93] Lavoisier's *Elements of Chemistry* clearly displayed this endeavor, for it began with a discussion of the cause of thermal phenomena, whether that cause was the imponderable fluid of heat, as Lavoisier believed, or a repulsive force, as Laplace argued. After this came a discussion of changes of state and the formation of gases in relation to heat, and only then did Lavoisier move to *physique particulière* in its more familiar chemical guise. When he did so, he presented caloric, the principle or element of heat, as one of the simple substances in nature. The phenomena of heat were attributable to a material thing, caloric, in Lavoisier's system, just as they were attributable to material phlogiston in his opponents' view. Thus, for example, James Watt wrote to Erasmus Darwin in January 1781, informing him that at the next meeting of the Lunar Society, "it is to be determined whether or not heat is a compound of phlogiston and empyreal air . . . what light is made of, and also how to make it." Humphry Davy, on the other hand, like his patron Count Rumford, had early opposed the materiality of heat. Davy in his "Essay on heat, light and the combinations of light" (1799), wrote admiringly of Lavoisier's system, whose sole defects seemed to be "the assumption of the imaginary fluid caloric, the total neglect of light." In the following year, William Herschel published an account of the heating effect of light and the refrangibility of heat, demonstrating an analogy between radiant heat and light. Newton, in the queries to the *Opticks*, had speculated about the interaction of heat, light, and matter. The variety of evidence, arguments, and counterarguments about the nature of heat and light that circulated in the early 1800s made for lively debate and a rash of further speculation.[94]

Schelling, in his *Ideen zu einer Philosophie der Natur* (Landshut, 1803), mentioned that several scientists "have considered light as a modification of heat. This view seems false, since not all heat can become light as every light can become heat." Coleridge asked in the margin whether all matter did not pass into light at a certain degree of heat. And in 1831 he asserted that "heat is the mesothesis or indifference of light and matter." Such notions, however, were based not on the assumption of the materiality of heat, but rather upon a philosophy in which matter was generated by powers, and in which heat was contemplated as power.[95]

Coleridge objected to talking of the cause of heat as caloric, for "caloric implies according to the use & analogy of language that the subject is a *thing*." But the word was established; so he worked with it, acknowledging Lavoisier's and Laplace's view of caloric, but suggesting another possibility, that caloric might be "a Quality, Property,

Function or particular modification of some higher Power." The production of heat in respiration seemed to him to provide evidence against the materiality of caloric, for there was no tenable mechanism whereby the nerves could acquire and transfer "the ubiquitarian *Thing*." Solid corpuscular philosophy was a sham and a humbug. Insight into nature would come through dynamic philosophy.[96]

Steffens in his *Grundzüge* developed the compass of nature as a quadruplicity of power, whose total unification or indifference constituted warmth or heat. But the north–south polarity in the compass was under the power of gravity, the west–east polarity being under the power of light. Light and gravitation, the primary powers in Coleridge's as in Steffens's cosmology, thus yielded heat as their indifference. In 1817, Coleridge adopted Steffens's view of warmth as the indifference of light and gravitation, and he maintained this view into the 1820s. In the language of powers, gravity, light, and warmth became "the three great coefficients of Nature."[97] Heat was important within dynamic philosophy, chemistry, and physiology. But philosophically and symbolically it was secondary to light.

Light and colors

The imaginative power of the divine injunction, "Let there be light," the symbolic role of light in Christian theology, and the philosophical and vital primacy of light in *Naturphilosophie* made it dramatically significant for Coleridge. It is striking that substantial entries on light and colors occur in his notebooks even when theological issues come to predominate among his concerns. Not all the sciences were so favored.

Coleridge became interested in light and colors before Goethe or Oken had published their books on the subject. Like them, he saw Newton's fragmentation of sunlight into the colors of the spectrum as totally false to the integral harmony and dynamism of light.

Newton, as he reported in his *Opticks*, had caused a narrow beam of light to fall on a prism by interposing between the prism and the sun a screen and blind in which slits had been cut. If a white card was suitably placed beyond the prism, then the beam of sunlight, after passing into the darkened room and through the prism, produced a series of colored bands on the white card. If one of these colors was isolated, by making a slit in the white card where it had appeared, then that color remained unchanged when it was passed through a second prism. There was another constant aspect to the phenomena, for the order in which the colors appeared on the first white screen was invariant, from red through orange, yellow, green, blue, and indigo, to violet. In addition, each color was bent to a different and characteristic extent on passing through a prism of a given medium, for example, flint glass. As we would say, each color in the spectrum had its own refractive index.

Newton's interpretation of these and related phenomena described in the first book of the *Opticks* was that white light was not homogeneous, unlike the seven colors of which it was composed. Each individual color in the assembly constituting white light was refracted to a different degree on passing from one medium to another, so that the assembly of white light was dispersed into its constituents by the prism. One consequence predicted by this theory was that images seen through lenses – for example, in using astronomical telescopes – would suffer from blurring because the component colors in the light passing from the body observed to the observer would be dispersed by the lenses. Newton believed that dispersion was directly proportional to the mean refraction of a given substance.[98]

His theory was not accepted by the Fellows of the Royal Society without lively debate, for it ran counter to the whole tradition of the theory of colors dating from Aristotle's discussion in *De Anima*. Aristotle wrote: "What is capable of taking on colour is what in itself is colourless as what can take on sound is what is soundless; what is colourless includes (a) what is transparent and (b) what is invisible or scarcely visible, i.e. what is 'dark.' The latter (b) is the same as what is transparent, when it is potentially, not of course when it is actually transparent; it is the same substance which is now darkness, now light."[99] Light and transparency were thus both necessary if color was to become visible; and potential transparency could be understood as darkness. Now in Newton's experiments, the colors were produced only when light and darkness were juxtaposed, the beam of sunlight passing into an otherwise darkened room. It was thus possible to argue that Newton's experiments had not invalidated Aristotelian explanations of color. Indeed, one could argue that by ignoring the active role of darkness in the production of color, Newton was being false to the empirical evidence before him. He was, in short, a sophist. Such a position was tenable only, if at all, when one ignored or rejected Newton's clear demonstration in book 1, part 2, that "the Phaenomena of Colours in refracted or reflected Light are not caused by new Modifications of the Light variously impress'd, according to the various Terminations of the Light and Shadow."[100] But there were those who did reject Newton's argument here. Theirs was the tradition, reinforced as we shall see by dynamic polar philosophy, to which Coleridge's criticisms of Newton belong.

He had long been interested in the theory of light and colors, attempting to perform experiments with a prism supplied by Tom Poole, writing to Poole in terms roundly abusive of Newton's theories, and making notes on the 1721 edition of the Opticks. "That a clear and sober Confutation of Newton's ~~Optics as far as~~ ⟨Theory of⟩ Colors ~~as concern~~, is practicable, the exceeding unsatisfied state, in which Sir I. Newton's first Book of Optics leaves my mind – strongly persuades me." Again, "I utterly reject the Newtonian Fiction of the solar light as consisting of similar fibres, each fibre consisting of seven dissimilar fibrils, and adopt the doctrine of Pythagoras [!] respecting Colors, as arising from Light and Shadow." How were the supposedly distinct colors brought into unity in a ray of light? How could the prism function as "a mere mechanic Dissector" of this ray? Such notions had "always, and years before I ever heard of [Goethe,] appeared monstrous Fictions!"[101] The key to the whole problem, beyond the necessary rejection of arid analysis, lay in the idea of darkness and its polar opposite, light. Coleridge spelled out his concerns in a note of around 1820:

> The distinction, however, between Black and Dark, if we employ one to express the contrary of, and the other the opposite to, White, is of great value. Among the numerous Sophisms and equivoques of the Newtonian Prism, the confusion of two senses in one term, Shadow, has not been the least fruitful of bastard notions – the first sense, the privation of light, the other, the relative diminution of the same by ⟨partial interception⟩. In the latter, the Shadow is *something* – a + seeing, in the former a nothing, a – seeing. – Now the Prism casts a Shadow in the latter sense, as a dense and semi-opake Whole; but it likewise casts 1. Color by its total *energy*, as qualifying (not intercepting) Light: i.e. the Prism generates *White* within itself – and 2. it casts *Colors* generated within itself by the polarizing energies of its parts acting on the White.
>
> Thus the Prismatic Spectrum is a highly complex phænomenon – so that, in the present state of our knowledge, the same appearance is susceptible of several Solutions.[102]

The best-known criticism of Newton's theory in Coleridge's day was Goethe's *Zur Farbenlehre* (Tübingen, 1810), learned in the qualitative though not the quantitative history of light and colors, and stressing the role of darkness with that of light in the production of color. Coleridge had written to Crabb Robinson in December 1812, asking him, apparently for a second time, and urgently, to procure a copy of Goethe's book: "In a thing, I have now on hand, it would be of *very important Service to me*." Goethe's theory was essentially conformable with Coleridge's dynamic philosophy; it was compatible with Steffens's cosmology and also with the seminal role of light and darkness in Genesis. Thus, Goethe became for Coleridge the modern spokesman of "the most ancient & pythagorean theory of Color."[103]

Goethe had examined Newton's *Opticks*, and noted that colors formed when light passed through a prism were associated by the displacement of "circumscribed objects . . . by refraction." This displacement, occurring over a dark boundary, was necessary for the appearance of color. Goethe argued that darkness was as much a principle as light was, that white was visible light, black was visible darkness, and color was the product of their interaction. He distinguished between interaction and merely quantitative mixture, the latter yielding only gray. Thus "colours are acts of light; its active and passive modifications." Goethe's theory was dynamic and qualitative, and so doubly attractive to Coleridge.[104]

Whereas Newton had been concerned with objective phenomena, Goethe's concern was with the subjective phenomena of perceived color. The first volume of *Zur Farbenlehre* accordingly began with a discussion of "physiological colours," which belong "to the *subject* – to the eye itself. They are the foundation of the whole doctrine." To him, afterimages and colored shadows were subjective. Goethe then discussed "physical colours," transmitted through a transparent medium. Their manifestation could be through reflection, light falling on a sharp edge, and other physical processes. The discussion of physical colors comprehended achromatism. In the mid-eighteenth century, it was shown that the relation proposed by Newton between diffraction and dispersion did not hold. John Dollond found that a combination of crown and flint glass could produce refraction, while virtually eliminating dispersion, one glass compensating for the dispersion produced by the other. Dollond applied the theory and built achromatic telescopes.[105]

Goethe saw in this discovery the beginnings of opposition to Newton's theory, and wrongly proposed that the distinction between refraction and dispersion should be abolished. Coleridge's perception of his own lack of understanding of this problem was clearer than Goethe's: "The difference between Refraction and Dispersion is not clear to me, as yet." Nor, in spite of his predilection for qualitative science, was Coleridge blind to the mathematical strengths of Newtonian optics. He remarked that Goethe, in spite of his experimental efforts to refute Newton, had himself confessed "that he had not succeeded in convincing or converting a single Mathematician, not even among his own friends and Intimates." Coleridge, himself no mathematician, wrote to Ludwig Tieck asking him what specific objections mathematicians had raised to Goethe's theory, "as far as it is an attack on the *assumptions* of Newton."[106]

Goethe's empirical arguments based on the subjective perception of color were forceful. Green, for example, could be obtained by mixing yellow and blue; yet Newton called it a pure color. "No artist," wrote Hegel, in a passionate defense of Goethe, "is stooge enough to be a Newtonian." For him, as for Goethe, color was color, however produced, and Newton's objective and analytical categories were simply irrelevant.[107] Here again, Coleridge, however lacking in understanding or sympathy for Newton's analysis, was willing to be persuaded by detailed evidence. The French chemist C. L. Berthollet had stated, in his *Éléments de l'art de la teinture* (Paris, 1791), that although many green dyes were produced by the mixture of blue and yellow, the green of plants was homogeneous. This struck Coleridge as "a very noticeable fact . . . the strongest fact, I have yet heard, in favour of the Newtonian Chromatology."[108] Coleridge, unlike Goethe, was concerned with light in its objective as well as subjective aspects, and he was therefore less blind to Newton's merits.

Coleridge respected Newton's experimental accuracy, while altogether rejecting the philosophical foundations of his explanations. An understanding of light and colors required a theory "adequate to the sum of the Phaenomena and grounded on more safe and solid principles." These principles would come from dynamic philosophy. "Goethe, & then Schelling and Steffens, had opposed to the Newtonian Optics [?the] ancient doctrine of Light and Shadow on the Grand principle of Polarity – Yellow being the positive, Blue the negative, Pole, ~~and~~ Red the Culmination ⟨and Green the Indifference:⟩."[109] Steffens in his *Grundzüge* had associated hydrogen and oxygen with the west and east poles in the compass of nature, and had especially identified the metals with the north–south axis. Oxidation and reduction could then be seen as deviations from that axis. But the west–east axis corresponded not only to hydrogen–oxygen, but also, and essentially, to the power of light. Colors, under the power of light, corresponded for Steffens to oxidation and reduction under the power of gravity, red being the oxidable or west pole, violet the reducible or west pole.[110]

Now Goethe had followed his section of physical colors by an account of chemical colors, colors fixed in bodies, ascribing their principal phenomena to the oxidation of metals. Steffens's theory helped to explain this. Coleridge, here seemingly indebted to Goethe and Steffens, wrote of color as light fixed in body.[111] The idea was to prove central in his schemes, for it posited a relation among light, color, and matter, which, translated into powers, meant that color was a product of the powers of light and gravitation. "Gravity in & subordinate to Light is color . . . ," Coleridge wrote in 1820; "I fear not to call . . . color the body of Light." Three years earlier, he had written to Tieck that "Color = Gravitation under the praepotence of Light," claiming "Behmen's Aurora" as the source for that idea. Whatever the inspiration, the formulation was Coleridge's own.[112] In a later scheme, color appeared as the synthesis in a standard Coleridgean pentad:[113]

	Prothesis Indistinction (the Mosaic Chaos)	
Thesis Gravity	Mesothesis Heat	Antithesis Light
	Synthesis Color	

Light and colors illustrated dynamic logic, and their total influence comprehended many different energies, including calorific and chemical ones.[114] The complex correspondences and interconnections that would become clear when light and colors were understood made them especially tantalizing to Coleridge, who in 1826 explained:

> Now the Problem is – to find some one ⟨Ideal⟩ Subject, which contemplated under the predicament of Multëity shall supply a Model or Canon, and therewith a nomenclature or scheme of Terms for the functions and affections of all other Subjects. And the most promising Subject for an experiment of this kind seems to me to be Light, as Light and as Color. – Preparatory to this, however, the several phænomena of Radiation, or rectilineal; of Scattering or hemispherical; of Reflection; Refraction; Absorption; Modification (the so called *Interference* of the Ray of Light); and of polarization; must be reduced to *Ideas*.[115]

There were many sources for learning the conventional meanings attached to technical terms. Their reduction to ideas required philosophy. Steffens had furnished valuable approaches to the problem. In 1808 the *Naturphilosoph* Lorenz Oken published a short work entitled *Erste Ideen zur Theorie des Lichts*. Coleridge, whose real interest in developing a theory of light dates from 1819–20, acquired a copy and worked through it in the early 1820s, experiencing as much frustration as illumination.

Oken began by postulating an ether pervading space, thrown into a state of tension caused by the sun and also depending on the planets. This polar tension was manifested as light, which traveled in straight lines because the tension was always between two bodies. Ether existed under two conditions, one being dynamically indifferent to body, the other having a dynamic relation to body; the former was dark, the latter light. When dynamic tension in the light-ether was changed by bodies, color or darkened light was produced. The more a body was colored, the more it had its own process of tension within itself. White and black on earth corresponded to light and darkness in the cosmos, true colors lying between them and determined by the degree of tension. Red was the noblest color, whose relations to white and black enabled one to define other colors. Yellow, for example, arose from the tension of red inclined to white, whereas blue arose from the same tension inclined to black. Or, in brief, yellow is a white red, and blue is a black red. Coleridge, already irritated at the style of Oken's criticism of Newton, replete with "rough Railing and *d——n-your-eyes-you-lie* Ipse-dixits," was further annoyed by the conceit and quackery of such infuriating nomenclature. But the scheme, however flawed, was dynamic, with yellow, blue, and red as the only true colors, from which all others were derived.[116]

Coleridge had earlier and unsuccessfully tried to derive a dynamic account of color, apparently from a consideration of the sequence of colors in Newton's rings as they were presented in the *Opticks*. He had also, again without success, tried to work from circles of colors like those described by Newton.[117] Now, more than a decade later, he tried again, working from Oken. Was Oken right? "Are Red, Blue, Yellow the three primitive Colors?" From this triad of colors, he derived a pentad, and from this in turn a heptad of seven colors: The heptad "was the largest possible formula for things finite, as the pentad is the smallest possible form." What was important, here as throughout physics, was the dynamic relation between colors.[118] It was this dynamism that underlay the relation of light and color to matter. Iodine, for example, was chemically electronegative and nonmetallic, and could therefore be expected to occupy a place in the compass of nature close to that of oxygen. But oxygen corresponded to the easterly power, and thus to darkness or negative light. Blue was the color that Oken and Coleridge after him had identified as inclined to black and away from red. Iodine gave a deep blue color when added to starch. Here was "a peep into the nature of Iodine as Light + Shade." Similarly, in 1819–20, Coleridge suggested that "the colors of different seas [arose] from different proportions of Iodine."[119]

In 1825–6, Coleridge tried to work out a more nearly complete treatment of color in relation to polarity and to matter. He proposed a heptad, with red as the "Zenith or Culminant Color," in which the powers "of Light and Shadow . . . exist in the maximum of energy." The nadir, in which darkness predominated, was represented by black. Yellow and blue were thesis and antithesis, indecomposable green was their indifference, and decomposable green the synthesis. White was a problem for Coleridge, and his notions about it shifted. The details of the heptad, and of his shifting views, are of less significance than his constant endeavor to distinguish the power of light from sensible light, and the "Oneness of the Power" of light, which "will appear as an ACT," from its "two polar forms or forces," which "will be understood as the *Material* factor."[120]

If color was derived from power, and if, as Coleridge believed, life was a power, there might be something to be learned about color from a consideration of its occurrence in life. Hunter had written *A Treatise on the Blood*, which Coleridge read. Everard Home, in his introduction, had pointed out that insect blood was without color. Higher organisms had colored blood. Why? Oken had concluded his multivolume *Lehrbuch der Naturgeschichte* by asking, "Why are there no green and blue men?" Coleridge addressed the question seriously, bringing together color and life as powers: "The reason may be – that the perfection of animal life consists in the continued antagonism of powers; but blue is the negative Pole of life, & marks deficiency – Green, the Synthesis, & where Synthesis should not be, *Confusion*, or Indifference – Bruise. Putridity." Everything *was* related to everything else.[121]

Sound

> O! the one Life within us and abroad,
> Which meets all motion and becomes its soul,
> A light in sound, a sound-like power in light,
> Rhythm in all thought, and joyance every where –[122]

Coleridge added this quatrain to "The Eolian Harp" in 1816–17. In the summer of 1817 he wrote to Tieck giving a dynamic account of the relation between light and sound: "Before my visit to Germany in

September, 1798, I had adopted (probably from Behmen's Aurora, which I had *conjured over* at School) the idea, that Sound was = Light under the praepotence of Gravitation, and Color = Gravitation under the praepotence of Light: and I have never seen reason to change my faith in this respect."[123] The language of powers and the primacy given to light and gravitation are clearly indebted to Schelling and, even more, to Steffens and to Genesis. The analogy between light and sound that Coleridge proposed was founded in dynamic philosophy and biblical exegesis, and may also have been strengthened for him by Thomas Young's discussion in the *Philosophical Transactions* of light, sound, and their analogies to one another.[124] Coleridge, developing a dynamic interpretation of the Mosaic account in a note of about 1820, stated that "light in and subordinate to Gravity is *Sound*," whereas gravity in and subordinate to light was color. "Hence," he observed, "the connection of sound with the hard & the Metallic. w*ch* are the especial exponents or representatives of Gravity in the world of the senses: and I fear not to call sound the soul of Gravitation and color the body of Light."[125] This note draws not only on Genesis, but also verses 1–3 of the Gospel of Saint John. But the Bible and the *Naturphilosophen* were not his inspiration for the account of sound. Coleridge was accurate in identifying Boehme's *Aurora* as his source.

Boehme followed the Hermetic tradition in seeing the tension between opposed powers as fundamental in the creation and progress of the world and soul.[126] He identified the union of powers manifested by God as the "Salitter," from which, through a succession of opposed powers, the world of matter and spirit was evolved. Here was Coleridge's "one Life within us and abroad."

Sound and light were for Boehme the two primary agents in God's creative work. Genesis began with the divine fiat, "Let there be light," and John presented that creation as enacted by sound: "In the beginning was the Word, and the Word was with God, and the Word was God." Words were indeed, as Coleridge claimed, "LIVING POWERS."

In *The Aurora*, Boehme wrote of "*two* things to be considered: *First* the *Salitter* or the Divine Powers, which are moving springing Powers . . . ; The *second* Form or Property of Heaven . . . is *Mercurius*, or the Sound, as in the *Salitter* of the Earth there is the Sound, whence there grows Gold, Silver, Copper, Iron, and the like; of which Men make all Manner of *Musical instruments*." Coleridge was most impressed, noting in the margin "§§27 is admirable – the Messenger or Mercury of the Salitter is indeed Sound, which is but Light under the paramountcy of Gravitation. It is the Mass-Light. The granite-blocks in the vale of Thebais shall send forth sweet *sounds* at the touch of Light – a proof that Granite is a metallic composition."[127]

Naturphilosophie, the system of creative and constitutive powers, the Bible, matter theory, the history of the earth, and Paracelsian and Hermetic ideas transmitted by Boehme have all come together in Coleridge's account.

6

GEOLOGY AND CHEMISTRY

THE INWARD POWERS OF MATTER

Beddoes and Darwin, Werner and Hutton

Coleridge, in the year before his death, had come to see geological debates as of the most fundamental importance. "Since the controversy between the Realists and Nominalists of the 13th, and 14th Centuries, there has been . . . no more important Question – than that . . . of the German & *thence* the French Hypothesis of a progressive Zoogony, (and of course, a Geogony, with its successive Epochs & their Catastrophes; and of its English opponents, Lyell . . . &c."[1] He had not always been so impressed with the importance of geology,[2] which he first encountered in the persons and writings of Darwin and Beddoes. Coleridge, disputing with Darwin in Derby in 1796, learned that he had adopted James Hutton's theory. Not long previously, Coleridge had read Darwin's *Botanic Garden*, which evinced a partly Wernerian stance toward the theory of the earth.[3] His encounter with Darwin came shortly after his first meeting with Beddoes, who was in the habit of addressing combative geological letters to his friend and fellow physician Darwin. Beddoes had studied at Edinburgh in the 1780s, when Hutton's theory was becoming known there. Beddoes's publications and lecture notes following his removal to Oxford show him as indebted to both Werner and Hutton. His lectures were essentially Wernerian, his papers Huttonian. Coleridge had certainly encountered the debate between the two schools of geology by 1796.[4]

Because Werner's theory remained important for German nature philosophers,[5] including Steffens, whereas Hutton's theory became the foundation upon which Lyell was to build, both theories were to be of enduring significance for Coleridge.

Abraham Gottlob Werner was the most influential teacher of mineralogy and geology of his day. Steffens, Coleridge's principal source in his construction of a chemical geogony, attended Werner's lectures in Freiberg, where he learned of his views about the classification of minerals. Werner believed that the external characteristics of minerals were necessarily correlated with chemical composition, although chemistry alone was inadequate in arriving at a classification of minerals.[6] He sought to unify mineralogy with historical geology, arguing that the surface of the earth had once been entirely submerged beneath the ocean, from which the rocks of the present crust had been deposited by sedimentation or precipitation. The sequence of deposition furnished both a history of the earth and a classification of its minerals.[7]

Werner's final system divided the formation of rocks into five periods. In the first or primitive period, rocks crystallized as chemical precipitates from a calm ocean. Granite was the first rock so formed. As the ocean became increasingly turbulent, the rocks formed were less crystalline. No life existed throughout this period. Then came a transition period, when life existed and the oceans were turbulent. The third or *floetz* period began with a shallow ocean, was stormy, and saw a rich development of life and another worldwide flood. Finally, and like the *floetz* period extending into the present, came the volcanic and alluvial periods, whose rocks were the result of local conditions rather than universal deposition from water. These five periods provided a framework within which to unravel stratigraphy as well as mineralogy. The relation between chemistry and mineralogy, the aquatic origin of most rocks, the unique status of granite as the most primitive rock, and the possibility of associating different stages in the development of life with different stages in the formation of the rocks of the earth's crust were all foci for contemporary debate, and all came to be significant for Coleridge.

James Hutton, a good friend of Beddoes's teacher Joseph Black, almost contemporaneously developed a very different theory of the earth.[8] Whereas Werner had directed his efforts to achieving a classification of the rocks, Hutton was more concerned to arrive at an understanding of the relation between changing landforms and unchanging geological processes. Werner's geology postulated unique periods and events. Hutton's envisaged an endless cycle, with no vestiges of a beginning, and with the same processes operating in nature now as had operated in the past. Werner saw rocks as deposited from solution. Hutton believed that heat and pressure were the principal agents in their formation, so that granite and basalt were of igneous origin. Beddoes and Darwin both gave much attention to the origin of granite and its relation to basalt, and Beddoes especially was eager to explore the power of chemistry in geology.

Coleridge, little interested in geology in the 1790s, would nevertheless remember these issues twenty years later, when Hutton's uniformitarianism would appeal to him less than the historical geology of Werner. The focus of geological debate had meanwhile shifted. Cuvier's studies of fossil remains had persuaded him of the reality of extinction, which resulted from catastrophic geological changes. As time advanced, new and higher species emerged.[9] Werner's theory of the earth, with its repeated inundations, could in principle be reconciled with catastrophism – hence Coleridge's association of "the German & *thence* the French Hypothesis of a progressive Zoogony" with "a Geogony, with its successive Epochs & their Catastrophes."

In 1799, Coleridge had talked with Greenough in Göttingen and walked with him in the Hartz. When Greenough in 1804 urged Coleridge to try Sicily – Catania and Mount Etna – for his health, geology was surely part of their conversation. Humphry Davy developed a knowledge of mineralogy as part of his work at the Royal Institution. Davy in 1807 was instrumental in founding the Geological Society of London, of which Greenough was the first president. It is striking that the early membership of the society also included Everard Home and Charles Hatchett, who with Davy formed the core of the Society for Animal Chemistry. And Brande, active among the animal chemists, lectured on geology at the Royal Institution in 1816, adopting many of Davy's views, and publishing the lectures in 1817 and again in his *Manual of Chemistry* in 1819; Coleridge studied this manual carefully.[10]

Dynamic geology, Steffens, and the *Theory of Life*

Coleridge was directly and indirectly kept aware of geological debate from 1799 to 1819. His real interest in geology, however, developed through his reading of nature philosophers, in an attempt to formulate the principles of a philosophical cosmogony and geogony. Oken, whose *Lehrbuch der Naturgeschichte* he was reading from about 1820 on, defined natural history and nature philosophy in a way that fully justified Coleridge's concern with geology: "Natural history is the history of the development of the planet plainly told, without any representation of causes. The natural history of the planet, when causally developed, is a part of nature philosophy."[11] Steffens was preeminently the writer who combined natural history with natural philosophy in investigating the history of the planet, so that here, as often elsewhere, Coleridge turned to him.

The first document evincing his geological indebtedness to Steffens was the *Theory of Life*. Coleridge there rejected theories that, like Bichat's, opposed life to death and severed the organic from the inorganic realm. Organism rather than mechanism was Coleridge's ruling metaphor, leading him to attempt to fill up "the arbitrary chasm between physics and physiology." Later, indeed, in 1826, he argued that there was such a chasm: "Evolution as contra-distinguished from opposition, or superinduction *ab aliunde*, is implied in the conception of LIFE: and is that which essentially differences a living fibre from a Thread of Asbestos." But in 1816 he saw the development of so-called inorganic nature as strictly a progressive, evolutionary, and correspondingly living process.[12] Coleridge here was not merely resting on *Naturphilosophie*, but also embodying concepts drawn from older Neoplatonic philosophy to arrive at his own interpretation of life in nature and the life of nature. The power of life worked at every level of nature, first in the union of powers and properties in gold and other noble metals, and secondly in the formation of crystals "as a union, not of powers only, but of parts." Then came

> the third step [which] is presented to us in those vast formations, the tracing of which generically would form the science of Geology, or its history in the strict sense of the word, even as their description and diagnostics constitute its preliminaries.
>
> Their claim to this rank I cannot here even attempt to support. It will be sufficient to explain my reason for having assigned it to them, by the avowal, that I regard them in a twofold point of view: 1st, as the residue and product of vegetable and animal life; 2d, as manifesting the tendencies of the Life of Nature to vegetation or animalization. And this process I believe – in one instance by the peat morasses of the northern, and in the other instance by the coral banks of the southern hemisphere – to be still connected with the present order of vegetable and animal Life, which constitute the fourth and last step in these wide and comprehensive divisions.[13]

Coleridge saw the history of the earth as a product of the power of life, the "Life of Nature," and at the same time as leading to the life of individual organisms, "the present order of vegetable and animal Life." He had a sense of process and symbiosis within, between, and throughout the realms of nature, so that the extreme margins and interfaces among those realms had a special fascination for him. Hence the excitement with which he contemplated coral and coal, mineral remains of animal and vegetable processes respectively: "The most pregnant historic Symbol on Earth is a Coral Bank on a Stratum of Coal – or rather a quarry of veinèd Marble on a Coal Stratum. The

Peat Moor and the Coral Bank, the Conjunctions copulative of animate and inanimate Nature! – Lime fertilizing Peat, and thus mutually effectuating each other's re-ascent into Life, the Peat into the nobler Gramina, the almost animalized Wheat, the shelly Lime thro' the Grasses into atmospheric red Life, and ending its brief cycle in Man!"[14] Chemical substances were transmuted through nature into the substance of living organisms. Geology thus drew on and led into chemistry, cosmology, and the life sciences.

More fundamentally in Coleridge's hierarchy of powers and sciences, geology manifested the ascent of power. The "original fluidity of the planet," the state necessarily antecedent to form, corresponded to the primeval chaos of Genesis, or the universal ocean of Werner's geogony. Then came successive polarizations, of which the first was symbolized by magnetism, and indeed by the earth's magnetic axis. In the compass of nature the north–south axis corresponded to the axis between powers symbolized by carbon and nitrogen, whose products were the metals. Here was Steffens's doctrine of the compass of nature applied to its limits. And what followed? "The metalleity, as the universal base of the planet, is a necessary deduction from the principles of the system."[15] The history of the earth then unfolded according to dynamic logic:

> From the first moment of the differential impulse – (the primæval chemical epoch of the Wernerian school) – when Nature, by the tranquil deposition of crystals, prepared, as it were, the fulcrum of her after-efforts, from this, her first, and in part *irrevocable*, self-contraction, we find, in each ensuing production, more and more tendency to independent existence in the increasing multitude of strata, and in the relics of the lowest orders, first of vegetable and then of animal life. In the schistous formations, which we must here assume as in great measure the residua of vegetable creations, that have sunk back into the universal Life, and in the later predominant calcareous masses, which are the *caput mortuum* of animalized existence, we ascend from the laws of attraction and repulsion, as united in gravity, to magnetism, electricity, and constructive power, till we arrive at the point representative of a new and far higher intensity. From this point flow, as in opposite directions, the two streams of vegetation and animalization.[16]

The compass of nature, with its elaborate correspondences, comes from Steffens's *Grundzüge*. There are, besides the *Grundzüge*, three other works by Steffens on which Coleridge drew in developing his ideas about geology. First, and briefest, is Steffens's essay on oxidation and reduction published in Schelling's *Zeitschrift fur spekulative Physik* (*1* [1800], 137–68). There Steffens first proposed a chemical approach to geology. Schelling, as Steffens noted on p. 143, had described vegetation or vegetative process as a constant reduction, in contrast to the continuous process of oxidation that was animalization. This made reasonable if superficial sense in view of the recent work of Priestley, Jean Senebier, and Jan Ingenhousz on the nature of animal and vegetable respiration. Steffens proposed to pursue this chemical hint in studying the living processes that he invoked to account for the history of the earth. He divided all minerals into combustibles, like diamond, sulphur, and metals, and bodies that had already suffered combustion, including earths, salts, and metal oxides (p. 149). Fire maintained constant chemical and therefore geological activity; the reduction of oxides was a process of vegetation, the combustion of sulphur or metals was one of animalization. Steffens had here achieved no more than a tentative correlation of Schelling's doctrine of organism with a chemical classification of substances significant in geology. He refined this somewhat (pp. 143–4) by invoking a general tendency to form crystals, an individualizing tendency that was to account for such phenomena as the occurrence of quartz crystals in granite. The explanatory model was, however, primitive and little developed.

Steffens made a significant advance in his next publication, *Beyträge zur innern Naturgeschichte der Erde* (Freyberg, 1801), a work that Coleridge read, annotated extensively, and used more than any other work in composing the *Theory of Life*.[17] The first section of the *Beyträge* is a long argument that carbon and nitrogen are the representatives of magnetism in chemical processes (pp. 1 ff.). The metals, products of northerly and southerly power conjointly, are then divided into two opposed series, one, the calcareous series, exhibiting the predominance of nitrogen or southerly magnetic power; the other, the siliceous series, exhibiting the predominance of northerly power (p. 8). Steffens then proceeds to support this division by a comparison of affinities and melting points, the more coherent or infusible metals being more akin to solid carbon, the less coherent ones being closer to gaseous nitrogen. The next step is to show that what holds true of laboratory experiments also holds for nature's laboratory, the earth, where Steffens seeks and finds a separation in nature of calcareous and siliceous rocks. Steffens notes that just such a separation is proposed in Werner's geognosy, where primitive granite is siliceous, and calcareous rocks are of later and separate formation (p. 15). The *Beyträge* continues with an elaborate chemical geognosy, substantiating the thesis.

As this geognosy is unfolded, Steffens remarks on the predominance of vegetable fossil remains in siliceous shales and of marine animal fossil remains in calcareous rocks, such as limestone (p. 22).

The correspondence is admirable: There are two chemical series, two geological series, and two routes pursued by nature in her organic operations. These routes, animalizing and vegetative, produce residues (p. 34). The silica in plants is educed by them, rather then produced by analysis. Siliceous rocks are the residues of nature's vegetative processes (pp. 38 ff.). Similarly, calcareous rocks are residues of nature's animalizing processes. Now, says Steffens (p. 57), silica arises from the carbon of plants by a living transmutation; and this carbon in turn arises from the siliceous row or series of mountains – notably granite mountains. Werner's placing of diamond with siliceous rocks strengthened this view. Thus carbon is the characteristic substance of this series, and it is also the characteristic element in vegetable nature. Certainly carbon is the principal element in plants. Similarly, lime is in the same series as ammonia, which contains nitrogen, so nitrogen is the characteristic substance of the calcareous series (p. 69). And nitrogen is a characteristic element in animal chemistry, and is comparatively lacking in the constitution of vegetables. Now the correspondences already developed can be extended (p. 89). There can be no doubt that in the whole calcareous formation we see only the tendency to animalization, while in the whole siliceous formation we see only the tendency to vegetation. Geogony must be considered organically, genetically (p. 93).

There follows a long discussion of the geographical distribution of metals, which, predictably, can be correlated with geomagnetism. The more coherent metals are commonest away from the equator, the less coherent ones near the equator (p. 174). The association of metals and minerals with magnetic polarity and with one another is pursued for another hundred pages.

The resulting brew was a heady one for Coleridge. Werner's geology, geomagnetism, the compass of nature, the life of nature, and dynamic chemistry all came together, imperfectly but attractively. Steffens in his theorizing showed an impressive acquaintance with the latest chemical analyses, geological explorations, and Werner's geogony and classification.[18] Dynamic philosophy and accurate data made a powerful combination, whose attractions for Coleridge are readily understandable. When he announced in the *Theory of Life* (p. 48) that the principal geological formations were "the residue and product of vegetable and animal life" and manifested "the tendencies of the Life of Nature to vegetation or animalization," he was adopting Steffens's principal conclusions as foundations for his own theory of the earth.

The last of Steffens's publications that is relevant to an understanding of Coleridge's approach to geology is his *Geognostisch-geologische*

Aufsätze (Hamburg, 1810), a work of which Coleridge particularly approved.[19] There Steffens pursued his earlier arguments, emphasizing the living polarities in geogony, developing his chemico-vital approach, and far transcending the limits of Werner's precipitation scheme (pp. 200 ff., 249). He now gave much attention to fossils, inferring from their nature and distribution that there had been major climatic changes. A note by Coleridge, surviving in transcription, explores these problems. He was struck by the tropical nature of fossils found in Siberia. "In order to [obtain] a full solution of this Problem," Coleridge concluded, "two Data are requisite; – First, a total change of Climate; and, secondly, that this change shall have been . . . instantaneous, and incompatible with the life and subsistency of the Animals and Vegetables in those High Latitudes and at that period and previously existing."[20] These conditions could be met by assuming a universal flood, followed by evaporation of the waters accompanied by a strong and steady wind. Rapid evaporation at the equator would, he believed, produce complementary cooling at "certain distances" from the poles – hence the barriers of ice in arctic and antarctic regions, surrounding, as he believed, open water at the poles. Geomagnetism was connected with climate; and the Bible gave authority for "a strong wind" that assuaged the waters of the deluge. Steffens went beyond Coleridge, arguing that climatic change had been accompanied, indeed caused, by inferred astronomical phenomena (pp. 290 ff.). The assumed nature and impact of these phenomena furnished the strongest arguments against uniformitarianism. Cosmology and astronomy were brought in as ancillary sciences in pursuing the study of the earth's crust. The examination of a single mineral raised questions that could lead one to confront the problem of the origin of the universe. Coleridge was more cautious here than Steffens, while remaining decidedly sympathetic to his program.[21]

Geology, chemistry, and the life of nature

The successive acts of creation, separated by "Relapses ⟨of Nature⟩ or Sinkings back from the organic and vivific [Tensions] commanded her by the word," were described in Genesis 1:9–25. Coleridge saw traces of these surges and relapses of power in the record of the rocks, where metallic ores were the "afterbirths" of creative acts, chronicled in Werner's classification of the successive formations of mountains. But if Werner's classification was right, his chemico-mechanical theory of the formation of rocks by precipitation was absurdly constricting, applying "to the plastic power of the Infinite a mete-wand borrowed from a chemist's Laboratory."[22]

The association of geology with chemistry was however, a fundamental one. Traditionally, mining, mineralogy, and metallurgy were closely connected with the theory of the earth. Werner had taught geognosy at the Mining School in Freyberg. In the early nineteenth century, the Board of Agriculture and the Royal Institution collaborated in carrying out chemical analyses of mineral samples, for the development of estates could involve the mining of minerals as well as the application of chemistry to agriculture. It is significant that Humphry Davy both analyzed minerals and lectured on soil chemistry for the Board. Brande, Davy's successor at the Royal Institution, also lectured on geology.[23] Steffens, following Werner, made chemistry the key to geology, and Coleridge, pursuing this approach, was especially alert for resonances between the two sciences.

From his first work on the *Theory of Life* until the early 1820s, Coleridge explored chemistry and geology. His cosmological and biblical enquiries led naturally into questions about the theory of the earth. But there is in Coleridge's approach to the sciences something of that series of surges and relapses, creation and quiescence, and constancy of productive tendency coupled with variety of productions that he attributed to the power of life in nature. Coleridge's world of nature was one decidedly built in his own image – necessarily so, because he was entirely convinced of the harmony between powers of mind and powers in nature. The character of his inquiry, the interdependence of his exploration of different sciences, of chemistry, zoology, and geology, is accurately displayed in a simultaneously retrospective and prospective notebook entry of 1820:

> Coleridgii Fides et Doctrina de Deo, Mundo, et Homine. Theosophy, Cosmogony Zöonomy and Anthropology. ⟨Fides Coleridgiana: or the⟩ System of Philosophy and philosophic Science taught in a series of Conversations during the years 1817–1820.
>
> [In two recent notebooks] I have flitted on, transcribing, criticizing, suggesting, side by side with Chemistry and her present Hierophants, Sir H. Davy, Dr Thomson and Mr Brande, ⟨above all, with Hatchett,⟩ thro' the Life of Nature in the organismus of our Planet to her potenziation by the . . . co-presence of *insulated* Life, both [vegetation and animal life,] and concluded with the Teeth and Bones. Now I recommence with Life [animal and organic?] queen-ing it in her own right, and *using* Chemistry, as one of her Hand maids, but Chemistry penetrated by her influence, transfigured and become vital. As I confine myself, however, to the actions and products, that have been watched, imitated, analysed by the Chemists and Galvanists of the existing Scottish and Anglo-gallican or Gallo-anglican School, the Subject will oc-

> cupy but a small number of pages – and then turning round vault back again to the Life and Exploits of Nature, geologically considered – wholly in order to collect the facts that have been observed and recorded by our Geologists and Mineralogists of best repute.[24]

Coleridge moved, although seldom linearly, from chemistry through cosmogony and zoogony to animal chemistry, then to zoology and physiology with chemistry in subordination, then back to geogony and geology – in order, as it transpires, to leap forward again into the life sciences.

One geological problem that he probed within this framework concerned granite, and was approached in notebook entries and readings concentrated in 1819–21.[25] Granite, as the primitive crystalline rock, was of fundamental significance. Thomas Beddoes had published an essay on granite in 1791, referring briefly to the existence of granite boulders in limestone mountains, and mentioning in a footnote H. N. de Saussure's observations on granite in his *Voyage dans les Alpes* . . . (4 vols, Neuchatel, 1779–96). Brande, in his *Outlines of Geology* . . . (London, 1817), and again in his *Manual of Chemistry* (1819), gave details of de Saussure's observations of granite blocks on limestone pillars near Mont Salève, and cited his assertion that the blocks had been swept down to their present position by great torrents when the Flood receded violently from the high Alps. The blocks had then protected the limestone beneath them from erosion, while all around the unprotected rock was worn down, leaving isolated limestone pillars capped by granite boulders.[26] Coleridge almost certainly read Beddoes, and may well have read de Saussure – and there were other sources in which he may have read accounts of these granite blocks, such as Playfair's *Huttonian Theory of the Earth* (Edinburgh, 1802, pp. 384 ff.). But Brande's version was the immediate source of a note in which Coleridge speculated that geochemistry might have produced the granite *in situ* by a reduction. He even identified the principal components of granite – quartz, feldspar, and mica – with powers whose relations might account for the succession of these minerals.[27] He always preferred a vital and dynamic explanation to a mechanical one. Brande, in contrast, Anglo-Gallican to the core, dismissed all speculative philosophy as Eastern allegory, considered Werner and Hutton as the only serious contenders in geological theory, lauded Davy's contributions to the science, and thoroughly aroused Coleridge's scorn.[28]

Brande was editor of the Royal Institution's *Quarterly Journal of Science, Literature, and the Arts*, which Coleridge read for facts that he then reinterpreted dynamically. An article by Henri Braconnot in

January 1820 on the conversion of ligneous matter into various chemical principles led Coleridge to reflect on the falling back of organic matter into minerals in the development of the earth's crust – metals akin to calcium from animal matter, metals akin to iron from vegetable matter. Metals were residues of organic action. This line of thought led him back to the arguments of Steffens's *Beyträge*, with its correlation among gravity, specific gravity or chemical force, distance from the poles, and the distribution of metals and peaty vegetable matter. Hence the following notebook entry of late 1819 or early 1820: "Q^{y}. – How far the revival of Metals may be attributed to decomposition vegetable? Whether at all or at all modified by the infinitesima of animal – medusas! Iron by gravitating Chemistry in cold Peat Climates! The ductile in hot climates, by sublimation! See Braconot's experiments on the substances formed from Wood by sulphuric Acid."[29]

Revolutions or uniformity?

The Bible, geological facts, and Coleridge's and Steffens's theories all agreed in accounting for the distribution of vegetable remains and metal ores with latitude.[30] Here was an attractive, tempting picture. And yet Coleridge believed that geology was *still* in its infancy. He happily dreamed up a story to account for the disappearance of the waters of the deluge. The hydrogen of the waters might have been transmuted, by a superinduction of power, into nitrogen, which, together with the oxygen in the waters, could then have passed into the atmosphere. Here was something to tackle Greenough with, if Coleridge, writing around 1822, ever met him again. "Where Knowledge = 0, Fancies that pretend to be no better than Fancies, are something, a sort of + 0 at least."[31] Coleridge generally knew how seriously to take his own speculative conclusions.

His ridicule of "our philosophising Noachists" was undoubtedly prompted by the controversy that followed the publication of William Buckland's *Vindiciae Geologicae; or the Connexion of Geology with Religion Explained* (Oxford, 1820). Buckland was ignoring the lesson that Galileo had taught, of the absurdity of reciprocally limiting and shackling science and religion. Granville Penn then made matters worse with his *Comparative Estimate of the Mineral and Mosaic Geologies* (London, 1822), in which he virtually treated Genesis as a textbook of geology, and complained against geology as an attack on revealed religion. Coleridge held forth to Crabb Robinson on June 3, 1824, on "the growing hypocrisy of the age and the determination of the higher classes in science to repress all liberality of speculation." He claimed that Davy had joined the party, "and they are now patronising Granville Penn's absurd attack on geology."[32] Coleridge, as we have seen, was given to exploring the scientific content of Genesis, but he did so with full regard for its literary and allegorical aspects; he did not seek to deduce geological science from revelation, but merely to explore their congruence. Granville Penn had arrived at his intolerant conclusion through an insensitive literalism that cramped thought. Freedom of scientific thought was as vital for Coleridge as the need to distinguish between fact and speculation.

Many even of the larger features of Coleridge's geological scheme were wholly speculative, although not contradicting the facts at his disposal. But there were some things that geology proved, chief among them being the genetic nature of the world. "Item," he wrote on August 27, 1823, "– the World not a total present, like a circle in space – but a manifest Spiral or infinite Helix in time & motion – Proved by Geology." Natural processes were progressive, and evidently so; witness, for example, the "Extinction of Volcanoes" and the production of islands from coral reefs. The progress, moreover, was convulsive. Coleridge wondered whether "every great Epoch whether of the physical or of the political or of the moral, World" did not end and begin "in a *Revolution* . . . – Over what petrified Tempest Billows must ⟨not⟩ the Geologist work his way in the attempt to penetrate beyond the products of . . . the present or human Epoch of this planet." His was an informed catastrophic view of the ascent of life. Nature, having set the animal form firmly on its feet in the elephant, dog, and horse, "proceeded to her next epoch-making Purpose, that of lifting it off its feet which she appears to have commence[d] quasi de novo in the Ornithorhyncus [duck-billed platypus], Tachyglossus [echidna], Kangaroo, Bat, Harpy, Galeopithecus [flying lemur], Maki [lemur], Baboon, Ape, Man –. But mark the interval" – the interval of the deluge, in which, "with a general *Sweep*, the noblest specimens of the noblest products alone *carried* forwards into the New Ledger of Life . . . "[33]

Coleridge's geology drew on Cuvier's catastrophism and was historical. In 1830 the first volume of Charles Lyell's *Principles of Geology* was published, in which uniformitarian principles were developed into a comprehensive system. Only those agencies currently operating in nature were assumed to have operated in the past, and Hutton's earth, showing no traces of a beginning, no prospect of an end, was Lyell's, too. Coleridge read G. Poulett Scrope's review of the volume in the *Quarterly Review*, and was predictably hostile to Lyell's approach. Uniformitarianism "applied to a world of growth & imperfection necessarily leaves a *craving void* for every philosophic Mind." Any account of the theory of the earth without beginning or end, he complained, "loses all pretention to the name of *Geology* – not to speak of the yet higher Geogony – it is merely *Geography*." Coleridge found Lyell's argument unsatisfactory because it took everything as simply given, instead of inquiring into its origins and production. For example, Lyell seemed to think that lime "must be furnished *for* the Coral-insect by submarine Volcanos – &c. How is this in the true spirit of that Baconian Logic, to which the English Chemists & Geologists make such proud appeals, against Werner & the German Theorists." Coleridge believed that coral generated the lime, although there might well then be a cycle "not unlike that of Yeast & beer / . . . And I do not see," he concluded, "why a Physiogonist should not be as much entitled to a conjectural origin in the mighty power of Life, as the Theorist to a conjectural volcanic power." In any case, said Coleridge in 1833: "Mr. Lyell's system of geology is just half the truth, and no more. He affirms a great deal that is true, and he denies a great deal which is equally true; which is the general characteristic of all systems not embracing the whole truth." Uniformitarianism, being founded on facts, was partly true – but it was false in its denial of the truths of "progressive Zoogony" or genetic science. In 1833, Coleridge, although favoring the progressive system, was unwilling wholly to reject Lyell's system. Instead, he anticipated "a third more comprehensive Scheme, which will contain & reconcile the truths of both . . . as a mesothesis."[34] Coleridge's openness to facts, his willingness to suspend judgment, and his insistence that the genetic aspect of nature must be considered led him more than once to hope for syntheses of opposed theories.

Atoms and elements

Chemistry was the key to Coleridge's dynamic geology. Davy's electrochemical researches in England, and Ritter's in Germany, had provided dramatic support for a new dynamism in science. Members of the Society for Animal Chemistry were pursuing their researches into the realm of physiology. Chemistry, which through Beddoes and Davy had provided Coleridge with his most stimulating introduction to the sciences, could be seen as at once the foundation and focus of the sciences, connecting the mineral world to the world of organized living beings and supporting scientific studies in both. Philosophically and empirically chemistry was exciting, dramatic, and popular, owing its vogue not a little to Davy's experiments and lectures. A letter from John Herschel to Charles Babbage in October 1813 captures the excitement: "I am glad to find you *agog* for chemistry. I am just now equally eager to prosecute that pursuit – and if Ryan continues in the same way of thinking, by the Lord, I think we may turn Peterhouse into a furnace, Trinity into a Laboratory . . . – I should like, as a first experiment, to make a party for breathing the nitrous oxide."[35] Coleridge, like Herschel, had dreamed of setting up a laboratory; his was to have been with Wordsworth and Davy. He studied chemical texts and papers more extensively and systematically than those of any other science. By 1810, and for more than a decade thereafter – throughout the years in which he read and used chemistry for his dynamic studies – chemists were engaged in the laborious but successful business of applying the teachings of Lavoisier, Dalton, and Berzelius to problems of inorganic and, increasingly, of organic analysis, determining atomic and equivalent weights and molecular formulas.

Lavoisier, by applying the concept of an element to express "the last point which analysis is capable of reaching," came to consider as elements "all the substances into which we are capable, by any means, to reduce bodies by decomposition." In his textbook of 1789, his table of "simple substances" listed thirty-three such elements.[36] Humphry Davy, who had eliminated some supposed elements from Lavoisier's list and discovered other new ones, in 1813 informed the reader of his *Elements of Agricultural Chemistry* (London, p. 8) that there were now forty-seven known "bodies incapable of decomposition." The number of elements was increasing steadily and unsatisfactorily for those who, like Davy, were convinced that nature was fundamentally simple.

Lavoisier's concept of an element was one pillar of early nineteenth-century chemistry. Another was John Dalton's atomic theory, in which different elements were characterized by atoms of different weights, all atoms of the same element – for example, gold – having the same weight. Once Dalton's theory became known, many chemists engaged in the determination of atomic weights. Dalton explained chemical combination as being between small whole numbers of atoms of different elements in simple and constant proportions. This explained the constant composition of most inorganic chemical substances, and enabled chemists to move from the results of gravimetric analysis to the construction of formulas exhibiting the number of atoms of each element in a molecule of a given compound. Even those who were skeptical about Dalton's atomic theory were mostly convinced of the validity of his laws of combining proportions, and investigated the equivalent or relative combining weights of elements.[37]

Many inorganic molecules seemed to display admirably the simplicity and constancy required by Dalton's laws. Organic molecules, however — molecules of substances from the realm of organized, living nature – were at first much more refractory for the theorist. They often contained large numbers of atoms, of which the commonest species was usually hydrogen, the element of lowest atomic weight. This meant that small quantitative errors in gravimetric analysis could result in significant errors in the molecular formulas of organic compounds. By 1815 chemists like Berzelius had begun to develop more accurate analytical techniques that enabled them to explore the validity of Dalton's laws for organic substances, and to produce some accurate organic analyses and formulas. It seemed that the same chemical laws might operate in both the organic and the inorganic realms.[38]

Thus, by the time Coleridge began systematically to try to apply dynamic philosophy to the natural sciences, chemists were busy breaking down all kinds of substances into their constituents, which they characterized by weight and number, by quantity, not quality, and which they identified with atoms whose aggregation constituted the formation of a chemical compound. There were criticisms of the program – Beddoes, Davy, and William Hyde Wollaston at different times and in different ways registered their objections – but chemists in 1815 were predominantly and clearly the heirs of Lavoisier and of Dalton. Their science was precisely of the kind that Coleridge abhorred.[39]

All his objections to mechanism therefore applied cogently to recent developments in chemistry, which he condemned as based upon what he regarded as intellectual fraud, distinguished by its "imposing simplicity" – imposing in every sense – and by poverty of abstraction.[40] How else could one describe a system founded on the teachings of Locke and of Condillac? Coleridge, with his constant stress upon unity and harmony, was directly opposed to Lavoisier's view that one learned about a compound by analysis. "The Chemist holds himself to know a Substance then first, when he has decomposed & by decomposition destroyed it, as *that* [Substance]. He understands only what he has mastered. Suppose, for instance, that a certain contagious miasma had Nitrogen or some Oxyd of Ammonium ~~plu~~ + Hydrogen for its constituent parts. By decomposing it, the now innocuous elements, ~~of~~ Ammonium and Hydrogen, would exist, but the *Contagion* would be no more." Chemistry should be a science of quality. But analysis, destroying qualities, could give no understanding of the nature of chemical compounds. Nor was this obscuring of knowledge produced merely by decomposition. Lavoisier saw chemical change as essentially the rearrangement of unchanging constituents, but such a view of the "Transposition ~~or inversion~~ of the Components, where two bodies are in question," could be uninformative. For example, Coleridge suggested, consider an exchange between calcium sulphate and sodium muriate [chloride]; "& we have Sulphate of Soda on the one side, while on the other instead of 'the Season of the Earth' we have a deadly poison, i.e. Muriate of Lime. – In other instances, the reversing or otherwise altering the *proportions*. Thus Oil becomes a mad'ning Spirit, the tasteless Water a corrosive Acid." In none of the cases were the original or the new qualities explained by Lavoisier's system.[41]

When Daltonian atomism was added to Lavoisier's transpositional scheme, metaphysical absurdity was twice compounded for Coleridge, who believed that it was all very well as long as one understood and employed the notion of atoms "as xyz in Algebra, and for the purpose of scientific Calculus, as in elemental Chemistry." As he wrote elsewhere, "An Atom is a Fiction of Science / or more properly, a symbol, like the plus and minus, . . . expressing operations of the mind, not representing *things* – subjective [as opposed to] objective. An atom expresses the unknown grounds of the union of the two opposite properties of Bodies, resistence or solidity, & divisibility. – To assert the reality of Atoms is to commence philosophy with a contradiction both in terms and in Sense." The contradictions involved in asserting that atoms were real "were exposed by Parmenides, 460 years A.C., so fully as to leave nothing to be added."[42] But many if not most contemporary chemists did believe that atoms were real, and they thereby aroused Coleridge's anger and frustration. Even Davy, whose galvanic researches surely favored dynamism, was unjustly convicted by Coleridge of the heinous crime of being an atomist. Brande, who came to the Royal Institution as Davy's successor in 1812, clearly considered that atoms were real. Coleridge apostrophized him in a scornful note: "Recoil, and retreat in good time, Man of Fashion – O Endive and Cellery of *European* and R.I. Philosophy – a single inlet of Light would disblanch your snowy Beauty . . . Right about to the Left and advance backward!"[43]

In July 1817 he had written to Lord Liverpool about principles, religion, politics, chemistry, and dynamic philosophy, arguing that "the late successful researches of the Chemists" almost forced on the very senses the fundamental truths of dynamism. Alas, chemists had relapsed to atomism in spite of the evidence before them. What was needed was clearly not atomism, nor, at the other extreme, dynamic schemes like Oersted's, which confounded matter with power alone, reducing phenomena and noumena to the same level of being. "The truth is, that neither the Atomic, or the Dynamic System, singly, can answer the demand," Coleridge observed in or around 1826, using "dynamic" in the restricted sense just indicated. "The matter is inactive without the Power . . . and on the other hand, the Power is objectless without the passive matter / . . . Both are but nomina abstracta, the one = agere, the other = pati. Why not then go at once to a reality, that contains the agere + pati, and agere = pati, + pati = agere, in its essential idea – i.e. to Life – first, as Unity and then under the condition of multeity (Lives)?" Coleridge's doctrine of powers was also one of life.[44]

Productive powers and chemical substances

Coleridge worked closely with his friend and disciple Joseph Henry Green in developing a scheme of vital dynamism during the 1820s. The application of dynamism to chemistry was Coleridge's, and predated the work with Green, but the fundamental nature of that dynamism was broadly constant throughout the 1820s. On September 21, 1830, Coleridge wrote that he considered that one great advantage of his and Green's philosophy was "that it establishes the distinction while it does away the heterogeneity of the Dynamic and the Material . . . Thus the dynamic (the Lebenskraft, for instance,) becomes actual in materializing itself; the Material, (the so called Stuff) vice versa is actualized in becoming a *live* body. Even chemical bodies are live bodies, in the dynamic moments tho' the flash-like Life not having found a cycle, they cannot be called *living* bodies."[45] Powers, matter, and life were dynamically related, although only powers and matter were of immediate concern when the system was applied to chemistry.

Before Davy fell from grace in Coleridge's eyes, he, with his galvanic researches, seemed fairly set to establish chemistry in the paths of dynamism. This was why Coleridge in his lectures on Shakespeare in 1811 told his audience that Davy "had reduced the art of Chemistry to a science, [and] . . . had discovered one common law applicable to the mind and body." Bishop Berkeley, likewise, critical of some aspects of mechanism and empiricism, and stressing the active role of mind, "needed only an entire instead of partial emancipation from the fetters of the mechanic philosophy to have enunciated all that is true and important in modern Chemistry." Coleridge even indulged in a dream of a priori chemistry, when powers of mind would suffice for a knowledge of the science. "Hereafter (and possibly the Time is not romantically distant) a chemical Newton may sit in his arm chair by his study Fire, and demonstrate the place, richness, and variations of a Mine in South America, in contradiction to the opinion of some Traveller who had been in the very spot – and a royal Commission upon examination shall confirm the verdict of the Speculator, whose Intellect was in his Eyes."[46]

Meanwhile, however, one needed philosophy and facts. Chemistry owed its fascination for Coleridge in no small measure to its blend of unifying ideas with detailed information about nature. For both personal and pedagogic reasons it was his preferred science. His clearest and finest statement of the imaginative power and significance of chemistry is given in a passage from the "Essays on the principles of method" in *The Friend* of 1818:

> the assumed indecomponible substances of the LABORATORY . . . are the symbols of elemntary powers, and the exponents of a law, which, as the root of all these powers, the chemical philosopher, whatever his theory may be, is instinctively labouring to extract. This instinct, again, is itself but the form, in which the idea, the mental Correlative of the law, first announces its incipient germination in his own mind: and hence proceeds the striving after unity of principle through all the diversity of forms, with a feeling resembling that which accompanies our endeavors to recollect a forgotten name; when we seem at once to have and not to have it; which the memory feels but cannot find. Thus, as "the lunatic, the lover, and the poet," suggest each other to Shakspeare's Theseus, as soon as his thoughts present him the ONE FORM, of which they are but varieties; so water and flame, the diamond, the charcoal, and the mantling champagne, with its ebullient sparkles, are convoked and fraternized by the theory of the chemist. This is, in truth, the first charm of chemistry, and the secret of the almost universal interest excited by its discoveries. The serious complacency which is afforded by the sense of truth, utility, permanence, and progression, blends with and enobles the exhilarating surprise and the pleasurable sting of curiosity, which accompany the propounding and the solving of an Enigma. It is the sense of a principle of connection given by the mind, and sanctioned by the correspondency of nature. Hence the strong hold which in all ages chemistry has had on the imagination. If in SHAKSPEARE we find nature idealized into poetry, through the creative power of a profound yet observant mediation, so through the meditative observation of a DAVY, a WOOLLASTON, or a HATCHETT;
>
> By some connatural force,
> Powerful at greatest distance to unite
> With secret amity things of like kind,

> we find poetry, as it were, substantiated and realized in nature: yet, nature itself disclosed to us GEMINAM *istam naturam, quæ fit et facit, et creat et creatur*, as at once the poet and the poem![47]

Chemistry, and indeed Coleridge's whole scheme of the sciences and of the "Genesis and ascending Scale of physical Powers, abstractly contemplated," symbolized relations of law, exemplified creativity in nature, and served pedagogically to call forth in the student

> the faculty of recognizing the same Idea or radical Thought in a number of Things and Terms which he had ⟨never⟩ previously considered as having any affinity or connection . . . Who ever attended a first course of Chemical Lectures, or read for the first time a Compendium of modern Chemistry (Lavoisier, Parkinson, Thomson, or Brande) without experiencing, even as a *sensation*, a sudden *enlargement & emancipation* of his Intellect, when the conviction first flashed upon him that the Flame of the Gas Light, and the River-Water were the very same things (= elements) and different only as A *uniting* with B, and AB united? or AB balanc*ing* and AB balanc*ed*?[48]

When Coleridge described chemical elements as "the assumed indecomponible substance of the LABORATORY" and "symbols of elementary powers," he was referring to the ideas expressed in the scheme of the compass of nature that he had derived from Steffens, for whom chemical elements symbolized powers corresponding to the points of the compass.[49] But Coleridge, unlike Steffens, insisted that the substances that chemists dealt with, carbon, nitrogen, and the rest, were bodies that had to be distinguished from the powers that they represented symbolically.[50] If the name of a body like carbon was used to refer to a power, it had to be understood that what was referred to *was* a power, so that carbon should be taken ideally. Thus it followed for Coleridge that "when a Philosopher shows me Carbon in its purest form of Diamond or Quarz, as the Body in which [the northerly power of attraction] is best represented . . . I know and understand what he means, and can give to the Ideal what subsists as Idea, and to the Real or ⟨Phænomenal,⟩ what exists as Phænomena." Bodies could be compounded with one another, but they were constituted by powers.[51] Now Steffens, for example, described the metals as composed of carbon and nitrogen, which were themselves apparently simple.[52] In the compass of nature, ideal carbon and nitrogen represented single poles and powers. But if they were taken as composing metals, they had to be taken as bodies, as things. Now, as Coleridge explained in the *Theory of Life* (p. 69), "nothing real does or can exist corresponding to either pole *exclusively*," because a thing was defined "as the synthesis of opposing energies. That a thing *is*, is owing to the co-inherence therein of any two powers." So it followed, as Coleridge noted in the margin of Steffens's *Beyträge* (p. 262), that a "simple Body is an absurdity: and if Carbon in its utmost purity and if Nitrogen in like manner be *Bodies*, they must each be composite no less than Gold or Arsenic . . . It is an error therefore and an inconsistency in Steffens to speak of the Metals as composed of Carbon and Nitrogen – unless where these are taken as the names of the Power predominant in each. And even so, yet not as composed of them, but as constituted by them."

Powers produced bodies: "The whole analogy of chemistry," Coleridge asserted in 1833, "proves that power produces mass." There remained the problem of chemical properties or qualities. Why were acids corrosive, and sugars sweet? Coleridge's answer involved two further concepts, coinherence and predominance. All four powers in the compass of nature coinhered productively in every body. What distinguished one body qualitatively from another was the proportion between these powers, and in particular which power was predominant. That relations between powers rather than bodies determined chemical qualities was apparent to Coleridge. The differences between diamond and charcoal, and between sapphire and clay, were trivial for the analytical chemist, who discovered the same so-called elements in qualitatively very different substances. There was carbon in diamond, charcoal, and even in carbon dioxide – here surely were proofs that bodies "have their essence wholly in *the Powers*." The theory of the chemist that "convoked and fraternized" superficially disparate bodies was founded for Coleridge on the doctrine of ideal powers and predominance.[53]

> There are four ideal Elements: pure Carbon, pure Azote [i.e., nitrogen], pure Oxygen, and pure Hydrogen, *ideally* indecomponible, and corresponding to these four elementary Bodies, ~~practically~~ indecomponible ⟨into bodies,⟩ each supposing ⟨all⟩ the 4 ideal Elements, but as under the predominance of some one of them. Thus Carbon would ⟨be⟩ C.H.A.O under the predominance of C. Azote = A.C.H.O under the pr. of A. Oxygen = OCHA, under the pr. of O. – Hydrogen, H.C.A.O. under the pred. of H – I.E. – they are practically indecomponible, because ~~eac~~ every one of the 4 Components in each is necessary to its being a body at all.
>
> To these may be added the Metals, as hitherto not decompounded and probably all *humanly* indecomponible, in the present epoch of Nature. Possibly, they are different proportions of Carbon and Azote indifferenced by the minimum of Hydrogen or Oxygen.
>
> All other bodies may be divided into ~~two classes~~ acids, Alcalies, and Salts; not as by chasms, but by degrees sufficiently different to constitute a technical diversity.[54]

Coleridge had so far restricted his account to the poles and center of the compass, and to the four elements and the metals corresponding to these points. The metals were principally distinguished from one another in this scheme by the relative predominance in them of ideal carbon and nitrogen. This was a modification of Steffens's classification of the metals into a coherent series, in which ideal carbon or fixity predominated, and an incoherent series, in which ideal nitrogen predominated. In both series, Steffens asserted, the specific gravity of the metals stood in inverse proportion to their coherence. Within each series, specific gravity thus indicated the specific chemical nature of the metals. Accordingly, specific gravity could be regarded as a chemical characteristic of bodies. Coleridge adopted this view, giving it his own exegesis. In the *Theory of Life*, he developed the doctrine of the three powers of length, breadth, and depth, identifying chemical power with depth: "That the *chemical process* acts in *depth* . . . is involved in the *term* composition . . . The alteration in the specific gravity of metals in their chemical amalgams . . . is *decisive* in the present [for the argument that amalgamation is a chemical change]; for gravity is the sole *inward* of inorganic bodies – it *constitutes* their depth."[55]

Steffens had observed that his law relating density to coherence was violated by platinum, an extremely dense and yet highly coherent body. Seizing on this clue, he had proposed in 1801 that platinum might be the last member of a hitherto unknown series or subseries of metals. Two years later, Wollaston discovered palladium and rhodium, and in 1804, Smithson Tennant discovered osmium and iridium. Coleridge, whose notebooks show his interest in these metals, regarded these discoveries as the realization of Steffens's prediction. The doctrine of powers appeared to be of demonstrable heuristic value.[56]

New elements for old: transmutation

The compass of nature was a fruitful symbol, even when restricted to its cardinal points. But, as Coleridge observed in 1818,

> it must not be forgotten, that the Compass of Nature like the Mariner's Compass is not designated by the 4 great Points, N.S.E. and W. only, but by the intermediates – Of these we ~~have~~ know that we have discovered two, the position of which is probably between E. and South, or Oxygen, and Azote – namely, Chlorine and Iodine. – It is possible that some known Bodies may have been confounded under Hydrogen, or some supposed decomponible – and others doubtless will be discovered. Oxygen itself may not be the due East, or Hydrogen due West–. /

> Chlorine and Iodine therefore must be taken as the representatives of the second Class of Bodies, the first ~~being~~ class supposing one predominant, and the other 3 balanced – while this second suppose a subordinate predominance likewise: if indeed it would not be still better to arrange the whole, as Powers, = 5 Forces = 10, Ideal Elements, Hypothetical elementary Bodies (or Chemical Tetrarchs) Existing elementary Bodies. Indecomponible Compounds – from Quartz to Phosphorus thro' all the series of Metals. – And lastly, ~~In~~decomponible Compounds, divided into Acids, Alcalies, and Salts. – And the transit from chemical to organic Bodies would be furnished by decomponible bodies chemically incomponible.[57]

Chlorine and iodine had first been characterized as elements in 1810 and 1813 by Humphry Davy,[58] whom Coleridge had subsequently unjustly convicted of the sin of subscribing to atomism. Coleridge regarded atomism as false and sterile, and hence incapable of leading Davy to his supposedly original discoveries. Steffens now replaced Davy in the pantheon of men of genius. "The demonstration that Soda, Potash, Lime, Barytes, &c are metals of the incoherent Series, given before 1800 by Steffens, were such that Davy's Exhibition of Sodeum [*sic*], Potassium, Calcium, Barium &c ⟨in 1809⟩ could scarcely increase our conviction." Coleridge became "most indignant at the continued plagiarisms of Sir H. Davy from the Discoveries of Steffens and others."[59] And as for the discovery of the elementary nature of chlorine, well! Coleridge remarked, "Long before Sir H. Davy's attempts to establish the independent existence of the Oxymuriatic as Chlorine, I had anticipated it a priori, tho' whether as an East by North, or East by South, I could not determine, but I conjectured the former. And such, I doubt not, it is."[60] Coleridge's dynamic classification of chlorine was supported by its "eager alloys" with metals,[61] suggesting a significant north–south component in the powers that produced it, and by its relative reluctance to combine with carbon, the representative of northerly power, because combination was most vigorous between substances in which opposite powers predominated. That Coleridge counted the formation of chlorides as one of alloying rather than of combination merely illustrates how theory can dictate interpretation. A similar spirit is apparent in his claim to have anticipated Davy's discoveries. Because Davy's researches on chlorine were completed some years before Coleridge's adoption, development, and application of the compass of nature, this claim is clearly false. But Coleridge surely believed it, convinced of the superiority of dynamism to the atomism he had rejected.

Coleridge's classification of chlorine and iodine identified them both with the easterly power in the compass, but modified toward the northerly power. This meant that the halogens should be electrochemically negative, like oxygen; and so they were. But it implied something more, and something that appeared startling, indeed impossible within the confines of conventional chemistry, the chemistry of Dalton and of Lavoisier. All real bodies were compounds, in that they were produced by all four cardinal powers in the compass of nature. They might nevertheless be chemically indecomposable, like Lavoisier's supposedly elementary bodies. Coleridge had developed the compass, so that a modification of power might produce a new chemically indecomposable substance. For example, the development of a subordinate predominance of northerly power, added to the predominance of easterly power, might, when the powers were appropriately balanced, yield the several halogens. Different bodies might be produced by the superinduction of different degrees and predominances of power. And this conceptual transformation might correspond to an actual transformation in the world of phenomena. In short, Coleridge's interpretation of dynamic chemistry made conceivable the alchemist's dream of transmutation, which remained an absurdity in atomistic chemistry.

Coleridge recognized this implication of his scheme, and sought evidence of transmutation in the writings of chemists whose own theories did not admit this possibility. The problems of contemporary chemistry in its applications to meteorology, agriculture, animal and vegetable physiology, and even organic analysis, were sufficiently complex for Brande, Davy, Daniel Ellis, and others, in spite of themselves, to furnish Coleridge with the evidence he sought.

Brande, in his *Manual of Chemistry* (p. 345), raised the question of the balance of gases in the atmosphere. Carbon dioxide was entering the atmosphere at an enormous rate, through combustion, fermentation, animal respiration, and other processes. Vegetables removed some of this gas, but Brande considered that their action alone did not suffice for the purification of the atmosphere. Coleridge, having in mind while reading Brande that the whole history of the earth manifested living processes, suggested that although plants were not enough, "yet by *vegetable* action and its analogy in the minerals & waters on the surface of the Earth the restoration may be as readily effected in the living atmosphere as in a living Man." How this might happen is hinted at in an earlier notebook entry, where Coleridge had considered that carbon might be capable of passing into oxygen "by change of state, the magn. for the electr.," northerly power being superinduced on easterly power.[62]

This speculation had arisen following Coleridge's consideration of

Davy's paper on a newly discovered explosive substance, nitrogen trichloride. Because chlorine occupied an easterly position i the compass of nature, and was now shown to combine with nitrogen, one might conclude that nitrogen, south ideally, had a westerly component or subpredominance in its constitution. Experiments by Davy, Berzelius, and others on the amalgamation of ammonia suggested that nitrogen might be compound. In one of his lectures at the Royal Institution, possibly in 1809, Davy asked: "May not Hydrogene and Nitrogene and Ammonia be all forms of the same species of Matter – combined or energetic in consequence of different electrical powers?"[63] Coleridge echoed the question in a notebook entry: "According to the dynamic View, Hydrogen is Ammonium in the state of + ~~Magnetism~~ Electricity, and Azote the same in the state of + Magnetism."[64] Elsewhere, and more succinctly, he wrote that "Hydrogen = elec$^{l.}$ Azot.: Azot = Magnetic Hydrogen."[65] Modern chemists were not the only ones to have suggested that there might be a common principle in hydrogen and nitrogen. Priestley, who had not grasped the relations between gases in the atmosphere and in solution, repeatedly purified water from dissolved gases, froze the water, melted the ice, and each time found as much nitrogen in the water as before. Here for Coleridge was confirmation of the apparent conversion to nitrogen of the hydrogen present in water. And, reading and rereading the works of Boehme, Coleridge repeatedly found prevision and confirmation of the transmutations later hinted at by experimental chemists.[66]

The possibility of transmutations involving water, hydrogen, oxygen, and nitrogen led to meteorology. Where did the water in the atmosphere come from when it rained? The Swiss chemist, geologist, and meteorologist Jean André Deluc waged an enduring fight against the neologues in chemistry and geology, the followers of Lavoisier and of Hutton, rejecting Hutton's view that water was dissolved in the air. Deluc's hygroscopic and meteorological observations indicated that there was just too much water during downpours to be retained in aerial solution. He argued instead that water was transformed into a different kind of gas in the atmosphere. And he supported this by citing experiments in which Martinus van Marum in Haarlem had decomposed water by an electric discharge, had noted the production of hydrogen, but had not noted the concomitant production of oxygen, which had combined with the wire bearing the discharge. Coleridge was delighted. If water could be converted by electricity to hydrogen, which could in turn be converted to nitrogen, then might not the reverse transmutational path also be open, from nitrogen to hydrogen to water?[67] Here was an explanation to baffle Daltonian chemists: "And our gallicizing Laboratory Atomists must explain the Phaenomena of Rain and the various results of Freezing more satisfactorily than they have hitherto done, before a verdict can be brought in against the superior Insight of" – Boehme![68]

Transmutation was associated with changes in the predominance of magnetic and electric powers, west – east being the electrical axis and north – south the magnetic axis in the compass of nature, here as always operating simultaneously at different levels of being and of meaning. Thunderstorms were known to involve dramatic variations in electrical power, Benjamin Franklin having shown that clouds changed back and forth between positive and negative electrical states during one burst of thunder.[69] One of Coleridge's notes records the observation that before a thunderstorm broke, the air was often heavy with a sulphurous smell – and the smell was stronger in places where lightning had just struck. Was the smell caused by sulphur, Coleridge wondered, and "if so, whence does the Sulphur come?" Instead of answering his question directly, he launched into a disquisition on the possibility that nature might be able to effect transmutations. He entertained similar conjectures in reading the account of the Aurora Borealis in Sir John Franklin's *Narrative of a Journey to the Shores of the Polar Sea* . . . The Aurora occurred in northern latitudes, affecting the compass needle though not greatly electrifying the air. Coleridge suggested that the phenomena accompanied a transition or transmutation, probably of magnetism – carbon and nitrogen – into electricity – oxygen and hydrogen.[70]

All this was at best speculative. But there appeared to be hard, precise evidence in the results of recent organic chemical analyses, which seemed to Coleridge to defy any explanation short of transmutation. Techniques of organic analysis were tolerably well developed and reliable by 1819 when Brande published his *Manual of Chemistry.* In one instance (pp. 362–3) he discussed the conversion of starch into sugar, giving the analyses published in Thomas Thomson's *Annals of Philosophy* (*6* [1815], 424–31):

	100 parts of Starch contain	100 parts of Starch Sugar contain
Carbon	45,39	37,29
Oxygen	48,31	55,87
Hydrogen	5,90	6,84
Nitrogen	0,40	0,00
	100,00	100,00

Coleridge read these figures, incorrectly, as indicating a transmutation of carbon into oxygen. He was on firmer ground in pointing to the disappearance of nitrogen. Brande believed that this was "no essential component," citing other analyses of starch in which no nitrogen was found. Coleridge, however, was legitimately of a different mind; the disappearance of an element suggested transmutation – "altogether most confirmative of my system."[71]

His eagerness to find support for transmutation and dynamic chemistry in the textbook of Anglo-Gallican Brande sometimes led him into error. Brande had given an account of simple hydrocarbons, including "olefiant gas" or ethylene (pp. 152–3). He then went on to discuss ethyl chloride, or "chloric ether," as Thomas Thomson had called it. Brande did so by transcribing some passages and compressing others from Thomson's popular *System of Chemistry*, a work from which he borrowed throughout the *Manual*, and which he with equal enthusiasm pilloried in reviews of its successive editions. In 1821 the review of the sixth edition in Brande's *Quarterly Journal of Science, Literature, and the Arts* condemned the book roundly: "There is scarcely a singular determination of Dr. Thomson's on any chemical subject of difficulty, during the last eight years, which has not been reversed." Thomson was, for the reviewer, incompetent, biased, confused, and a plagiarist.[72]

Brande nevertheless used Thomson's *System*, often without acknowledgment. He compressed Thomson's discussion of olefiant gas and chloric ether so thoroughly as to be misleading, concluding one paragraph with the statement that "this variety of carburetted hydrogen has been termed *olefiant gas*." The next paragraph began: "*Chloric ether* is the term applied to this fluid by Dr. Thomson." Not unreasonably, Coleridge assumed that Brande was saying that chloric ether and olefiant gas were the same. On the same page of his *Manual* (p. 153), Brande gave an analysis of chloric ether, which contained chlorine, carbon, and hydrogen. And on p. 416, he gave an analysis of ether into carbon, hydrogen, and oxygen, equivalent to specified proportions of olefiant gas and water. In these analyses, there was no chlorine, and more oxygen than the water alone would have contributed – or so it seemed to Coleridge, who was by now thoroughly muddled by Brande's inept plagiarism and by his own inability to cope with quantitative results. He dived into the morass and came up with the conclusion that "Olefiant Gas being chloric Ether" – which Brande seemed to claim but had not meant to, and which in any case was not true – it followed that Brande had described a transmutation of chlorine into oxygen – which he had not. Finally, as if inspired to compound Coleridge's already fine confusion, Brande had claimed that none of the chemical processes he had just described was ever perfect (p. 417). This must have seemed like a weak attempt to cover up evidence of transmutation, and it led Coleridge to his satisfied conclusion: "This, however, sufficiently favors my notion, & gives a *sense* to my queries & suggestions, written before I had come to this passage."[73]

Coleridge's selection of chemical evidence of transmutation appears so far to have been based on a truly stratigraphical succession of errors. There were, however, areas where the crudeness of contemporary chemistry resulted in controversial and even paradoxical results, which could properly be used to favor dynamism. What is more, Coleridge's sense of the logic of chemistry led him repeatedly to these paradoxes, often identifying crucial problems that were not to be resolved for another half century or more. Such problems were concentrated at the interfaces that so fascinated Coleridge, where chemistry was applied to physiology and to the chemical products of living organisms.

Coleridge was convinced that dynamism was the key to chemistry, alchemy its goal. He held to this conviction. In 1832 he informed his listeners that "alchemy is the theoretic end of chemistry: there must be a common law, upon which all can become each and each all." Coleridge, grown beyond his early disappointment at Davy's absorption by the establishment, looked back on him in his prime as "the Father and Founder of philosophic Alchemy, the Man who *born* a Poet first converted Poetry into Science and *realized* what few men possessed Genius enough to fancy." Davy had done this by developing from his galvanic researches a new theory of affinity and of chemical combination.[74]

Affinity and combination

Chemical affinity was the concept traditionally used for relating substances according to their reactivity. If substance C displaced B from its combination with A, $AB + C$ yielding $AC + B$, then C was said to have a greater affinity for A than B had. A principal goal of chemistry in the Enlightenment had been the interrelation of all known substances in a sequence of displacement reactions, and the establishment of a corresponding table of elective affinities.[75]

The program was attractive but simplistic. As chemistry became increasingly integrated with the natural philosophy of physicists in the late eighteenth century, the range and complexity of factors bearing upon chemical combination began to emerge. Lavoisier excluded consideration of affinity from his *Elements of Chemistry*, because this would have taken him into the transcendental part of chemistry. C. L. Berthollet tackled the problem in substantial monographs on the laws of chemical affinity and chemical statics. The result of such work was the destruction of hopes of organizing chemistry according to affinities. By 1800, when Jean Senebier published in his *Physiologie végétale* the claim that the history even of organized bodies would be known as soon as one had a general theory of the affinities between their component substances, he was espousing an approach recognized as overambitious and unfruitful.[76]

The year 1800 also saw the invention of the voltaic pile, which led to a host of galvanic researches, and to the development of electrical theories of affinity by Davy, Berzelius, Ritter, and others. Chemical action generated electricity, which, in turn, when passed through a wide range of substances, produced further chemical action. What was more, bodies could be arranged in electrochemical series. Their electrical relation to the galvanic pile or battery could be correlated with their chemical behavior. Perhaps galvanic or electrochemical phenomena could be explained by postulating either a common cause for or an identity between electrical power and chemical affinity. Davy considered both possibilities. The Enlightenment dream of organizing chemistry through affinities was revived, but with a difference, for electricity had been used, notably by Davy, to decompose formerly indecomposable substances, thus offering a new route into the secrets of matter. Abernethy argued that Davy's experiments and the rule of analogy combined to suggest that in living bodies something like electricty might account for organic change.[77] The new theory threatened to be as simplistic and as overambitious as the old.

It had, however, strong attractions for Schelling and his followers, for electricity was a polar power, the galvanic pile had poles, and an electrochemical theory of matter allowed proponents of the polar dynamic philosophy to incorporate chemistry into its scheme; Schelling did so in his account of the construction of matter.

Steffens took up and developed Schelling's account. Chemical combination was a process of achieving an indifference or dynamic synthesis. Water, for Daltonian chemists, was composed of oygen and hydrogen. Steffens defined it as the indifference of the relative difference of oxygen and hydrogen, electronegative and electropositive respectively. These electrical characteristics were the foundations of chemical activity in nature. Because Steffens saw each chemical process as endowed with its own characteristic individual life, manifested as a tendency to assimilation, the chemical activity of nature, founded upon electrical polarity, could also be seen as underlying the life of nature. Because more than the electrical character of bodies was determined by their place in the compass of nature, Steffens sought to correlate affinity with other properties, such as solubility, fusibility, and coherence.[78]

Coleridge explored these attempted correlations carefully, proposing that "Depth = Gravitation = Galvanism = Chemical Combination!" Or perhaps galvanism might be added to chemical affinity. The power of galvanism, exerted in chemical combination, was chemismus or constructive chemical affinity.[79] Coleridge stressed, however, to a degree that Schelling and Steffens had not, the dynamic nature of affinity, produced by the synthesis of electricity and magnetism. These powers, "contemplated as Acts of a Self-constructive Power, and identified in this Power," constituted chemismus. Galvanism, sometimes identified with chemismus, was at other times viewed as "the transition of Electricity into Chemismus or the co-adunation of Magnetism and Electricity." In short, Coleridge saw galvanism and chemical power as alike produced by magnetism and electricity, while remaining undecided about their identity – an indecision shared by Davy. What was certain was that active powers were essential to the very being of chemical combinations. The characterizing predominance of powers developed for an understanding of chemical elements was no static balance but an active and dynamic one, working equally in elements and compounds.[80]

When a body was chemically constituted, powers in polar opposition produced an indifference. Water, for example, was "the neutral product of oxygen and hydrogen." It was important to recognize the integrity of the synthesis. Water was not a mixture of two gases, but a synthesis, no less "a *simple* Body than either of the imaginary Elements, improperly called its Ingredients or Components."[81] Similarly, wine was not merely hydrogen, carbon, and oxygen, or alcohol, sugared water, and malic acid, but rather a combination "in a living process." The interpenetration of opposites produced something qualitatively new. In chemistry, "no co-existence is permitted[;] the powers destroy each other in the act of becoming some new power, which may either manifest itself as a positive power or as a copula of properties derived from the entire suspension what we might perhaps venture to call an intussusceptive equilibrium or chemical neutralization."[82]

This metaphysical account of what happened when two substances were brought together and combined chemically to produce another substance was unintelligible, beyond the reach of the understanding. How could galvanism, depth, gravitation, and chemical combination be all one? How could carbon, hydrogen, and oxygen combine to form wine? Coleridge knew that such unities and chemical changes could only be described metaphorically and grasped imaginatively. When he wanted to explain the difference between imagination and fancy, he contrasted juxtaposition with chemical synthesis, concluding that "the Imagination is the synthetic Power."[83]

Oxidation, combustion, and acidity

The use of the compass of nature to explain chemical combination was an early nineteenth-century invention. But it may not be altogether fanciful to see an ancient lineage behind it. Aristotle, like the *Naturphilosophen*, classified the fundamental qualities inhering in bodies into two pairs of opposites; he chose hot and cold, wet and dry. Varying proportions of these qualities characterized his four elements, fire (hot and dry), air (hot and wet), earth (cold and dry), and water (cold and wet). All four elements were present in all real sublunary bodies, whose qualities arose from the proportions among the elements. The elements existed in intimate association. Changes in bodies were brought about by changes in the elemental mixture.

The doctrine of opposites was preserved, while richly adulterated, by matter theorists and alchemists of the Arab and Latin West throughout the Middle Ages. Paracelsus, iconoclastic Renaissance physician, eliminated air from his list of elements, proposing instead his own *tria prima*, salt, sulphur, and mercury, embodying the solid earthy principle, the principle of combustibility, and the principle of fluidity – a simple revision of Aristotle's earth, fire, and water. Paracelsian views persevered into the eighteenth century, notably in the German states. The phlogiston theory of J. J. Beccher and G. E. Stahl was truly a successor of Paracelsus's chemical philosophy; phlogiston itself, the principle of inflammability, played a similar role to the sulphur of iatrochemists and alchemists.[84] When a body burned, phlogiston was released to the air. This theory was entirely chemical, relating to the qualitative changes of bodies and ignoring quantitative ones.

Lavoisier, by making chemistry part of a broad natural philosophy that included physics, made the measurement of weight central to the new chemistry and, much concerned with changes of state – solid, liquid, and gaseous – arrived at a theory of combustion that essentially inverted the old one. He showed that when a body burned, it took something out of the air. That something, when combined with, say, phosphorus or sulphur, yielded, with water, an acid. Lavoisier accordingly named it oxygen, the acid-generating substance or principle.[85] Thus oxygen, involved in combustion and apparently essential to acidity, came to appear as the most important element in Lavoisier's system – a role reinforced by its part in the chemistry of respiration. Lavoisier was so confident that oxygen was the acidifying principle that when he came to the acid obtained from sea salt and failed to find oxygen in it, he concluded that it was composed of hydrogen and a compound radical, part of which was oxygen, the other part being as yet unknown.

Combustion and processes involving heat had all along been at the heart of chemistry, from the days of the alchemists with their furnaces until Lavoisier's chemical revolution. The association through oxygen of combustion with acidity, respiration, and the rusting of metals conferred an imposing unity on chemistry – and aroused Coleridge's criticism. During the early years of the nineteenth century evidence was adduced that weakened this unity, in the cases of chlorine and of hydrogen sulphide. A century earlier, chemists had failed to appreciate Mayow's account of nitro-aerial particles, a part of the air active in respiration and combustion, because they generally thought of air as a simple substance and ignored the specifically chemical activity of its components. More recently, Coleridge believed, the same kind of error had been repeated by "Lavoisier and the French Chemists. Again the common property of several substances was substantiated under the name, Oxygen, into the constant principle of Acidity, flame, rust and life, – one and the same body in all instances."[86] Then Lavoisier's theory of acidity had been overthrown by Davy's demonstration that there was no oxygen in the acid from sea salt. "Hence," Coleridge concluded, "the importance to Science of the Distinction established in my 'Philosophy of Method[']' between Ideal Powers and the chemical Stuffs." He developed his own chemical symbols, using . . . to indicate the easterly power generally, including chlorine, iodine, and oxygen. "But Oxygen as a *Stuff* or Element of the Laboratory is signified by . ẋ̣ ."; elementary chlorine was . ch̤ . , the predominance of power differing in different stuffs.[87]

Davy had at first believed that the acid from sea salt did contain oxygen. Analogy made it probable. But his efforts to pull oxygen out of the supposed radical repeatedly failed. In the end, he was forced to the conclusion that the acid was not an oxy-acid, but rather a compound of hydrogen with an elementary body. Because this body was a green gas, Davy, anxious to avoid the pitfalls of theory-laden nomenclature, simply called it chlorine.

Davy had announced the elementary nature of chlorine in 1810, thereby rendering Lavoisier's theory of acidity untenable. Coleridge in 1812 used its overthrow to show how a single exception could destroy a theory. Perhaps because of his disenchantment with Davy, Coleridge ignored chlorine, instead observing that water was not an acid, although it contained oxygen, whereas hydrogen sulphide was an acid, although it contained no oxygen. It followed that "this whole Theory of Acids, as Oxides, is as completely subverted as if a thousand facts had been adduced." Later, when he had developed his account of qualities produced by powers, he saw hydrogen sulphide, like water, as an indifference, acidic with respect to strong alkalis, alkaline with respect to strong acids.[88]

The use of principles embodying functions – for example, phlogiston as the principle of inflammability or oxygen as the principle of acidity – was too firmly entrenched in chemistry to be rejected along with Lavoisier's theory of acidity. After Davy's early papers on the elementary nature of the halogens, his exact contemporary Joseph-Louis Gay-Lussac published a paper on acidity and alkalinity, salvaging what he could of the old theory and even extending it. All bodies were neutral, acidic, or alkaline. Oxygen was acidic, although one could no longer regard it as the unique principle of acidity, because an acid was simply a substance capable of neutralizing alkalis. Oxygen being acidic and water neutral, it followed that hydrogen, water's other constituent, must be alkaline. Davy was indignant at what he saw as Gay-Lussac's attempt to elevate hydrogen into an alkalizing principle. The chemical properties of compounds arose through combination, not through the mere addition of the properties of their constituents.[89]

This was a view perhaps inspired by Coleridge, and certainly shared by him. It invalidated those labels incorporating "the *gen* of French nomenclature," namely, hydrogen and oxygen. Coleridge early proposed using suffixes based on "archè" for "that which causes a compound to begin to be; and to express the Base, by combining with which it begins that compound, by the affix 'yl,' from υλη, the Stuff, or material Substrate" – a proposal that led to such monstrosities as "Hydroxalkalarchico – Carbonic Gas! Hyperhydroxalkalchated Muriatic Gas," relinquished even by their author.[90]

With the rejection of French theory and nomenclature, Coleridge had to look elsewhere for principles that would bring sense to chemical classification. What was an acid? Was it even philosophically justifiable to consider acids as a class?[91] German schemes, stressing combustion and susceptible of integration with the compass of nature, took him away from the problem of acidity and back to a chemistry of powers. First fire, then the sulphur of the alchemists, then phlogiston, and finally oxygen had kept combustion to the fore in chemical theory. In 1812, Oersted, in his book on the chemical laws of nature, divided bodies chemically into combustive, combustible, and neutral ones, or acids, alkalis, and salts, all deriving their properties from the universal forces of combustion and combustibility. Coleridge accepted that the electric poles constituted "the comburent or combustive." He expressed this as a triad, "which dynamically contemplated is

Combustive ♓ Combustible

Combust:

and contemplated as bodies, 1. Oxygen (+ Chlorine + Iodine). 2. The Metals + Hydrogen. 3. Oxyds, Acids, Alkalies, and Neutral Salts." And the polar opposites within the triad could also be interpreted as west and east in the compass of nature.[92]

Chemistry become vital

Coleridge's renewed study of chemistry beginning in about 1819–20 was undertaken within the wide context of a study of life – the life of nature, and life within nature. Davy, Thomson, Brande, and Hatchett were the authors whom he particularly acknowledged.[93] But in spite of Davy's role in his intellectual development, Coleridge worked most systematically and extensively with Brande's *Manual of Chemistry* and with the Royal Institution's house journal, known simply as "Brande's *Journal*." It was not merely the self-confident accuracy of Brande's facts, nor even his habit of publishing accounts of the very latest work in chemistry, that made the writings of this "scientific version of a Dickensian Gradgrind" so important to Coleridge. Medicine was central to Brande's professional life, and he "saw his goal as one of deepening the relationship between chemistry and medicine. Most of the articles he published during his lifetime dealt with some aspect of that grey area between physiology and organic chemistry." Brande's journal published extensive medical analyses; his *Manual of Chemistry* was designed for medical students who attended his chemistry lectures at the Royal Institution. His account in the *Manual* of animal and vegetable substances was full and comprehensive, and correspondingly serviceable to Coleridge, however antipathetic the underlying philosophy.[94]

When Coleridge in 1819 began his study of Brande's *Manual*, he was closer to Steffens and the *Naturphilosophen* than he later became, willing to consider chemical processes as animalizing and vegetative ones. This did not imply a naive reductionism. The successive levels in Coleridge's schemes of powers were distinct, so that whereas the vital realm comprehended the chemical one, life could not be explained by chemistry. But Coleridge nonetheless saw continuity as well as distinction throughout the ascent of nature. Even a pure metal or a crystal manifested life for him. This notion of continuity and transition in ascent enabled him in 1818–20 to explore the relations between chemistry and life. "The transit from chemical to organic Bodies would be furnished by decomponible bodies chemically incomponible."[95]

Within a few years, Coleridge came to see the distinction between the powers of chemistry and those of life as absolute. Formerly he had illustrated the power of life, the tendency to individuation, with the phenomena of crystallization as well as those of anatomy and physiology. But in 1823, repeating that "life is a tendency to individualize," he introduced a distinction between life and the "tendency to specificate or become an aggregate by apposition," as manifested by crystals, which grew without "evolution (*ab intra*)." A decade later, he argued that chemically indecomposable substances or "*Stuffs*" limited chemical science, keeping it in a closed circle with dynamics and statics. There was a "complete saltus" between these sciences and the life sciences.[96]

This fundamental shift in Coleridge's view of the relation between chemistry and physiology was part of his recognition and rejection of the implication of Schelling's view of nature as absolute and self-determined. Living nature was, however, productive and ever evolving. It was therefore of particular interest to Coleridge to see how far productivity operated in chemistry. The crucial area for this inquiry was at the proximate limits of chemistry and life, between physiology and organic chemistry, where Brande published most of his papers. Chemistry's power of elucidating problems in that area was at best doubtful in the early decades of the nineteenth century. It was one thing to analyze wine or bile – and even such analyses were difficult and uncertain in 1819 – but quite another to use chemistry to explore living processes like digestion. There was considerable force in Bichat's contention that organic analysis was "the dead anatomy of the fluids, not a physiological chemistry," and that there was an immense gap between the laws of physics and chemistry and the laws of life. Coleridge, whose initial opposition to Bichat was total, came to accept that there was a gap between the life sciences and the rest.[97]

Products and educts

Chemists in the late eighteenth and early nineteenth centuries were much occupied with the question of preformation, especially when considering complex organic materials. When, for example, the stem of a plant was analyzed, several distinct substances were isolated and characterized. Had these substances preexisted in the plant, to be separated but not essentially changed by analysis? Many chemists among Coleridge's Anglo-Gallican fraternity answered yes; Senebier even believed that vegetable chemistry would be capable of complete elucidation when the affinities of the relevant constituents were known.[98]

But was this true? Chemistry was not sufficiently advanced to decide, and in Coleridge's day chemists allowed for the possibility that preformation might not always work by distinguishing between products and educts. Watts's *Dictionary of Chemistry* defined "educt" as "a body separated by the decomposition of another in which it previously existed as such, in contradistinction to product, which denotes a compound not previously existing, but formed during the decomposition."[99] The distinction, clear in theory, was not always so in practice. For example, Brande noted that "Thenard separated from bile a peculiar substance, which he has termed *picromel*; but the process by which he obtained it, is so complex, that I think it doubtful whether it be a product or an educt." Coleridge, reading this passage, observed that it was matter of regret that Brande "had not devoted a chapter to the nature of this important Distinction, and the rules for its application. The Dynamists, much more the Zöodynamists, would regard most of his own Products, as Educts." Steffens made the same point, complaining that French chemists seemed to think that vegetation could be handled as a chemical process. But what they found in plants were not products but educts. Plants were composed of living parts – sap, roots, sexual organs – and residues – wood, rind – and they produced substances such as nectar. Steffens observed that chemical analysis told us about these products, but could not discover how the substances had been combined prior to analysis, because the laws of chemical affinity were transcended in living organized beings.[100] The distinction between educts and products was thus important, and it bore directly on Coleridge's concern with the issue of productivity in chemistry. He regarded so-called elements as in fact compounds, the stuff carbon being constituted by the powers carbon, nitrogen, oxygen, and hydrogen under the predominance of carbon, and similarly for other elements. The distinction between product and educt was therefore applicable to Lavoisier's elements as well as to his compounds. Could the results of analysis then reveal what had existed in a plant? More importantly for Coleridge, could a knowledge of the substances that went into a plant, through the roots and leaves, account for the substances found by subsequent analyses? The state of chemical knowledge in 1819 was such that the answer was frequently no. Nitrogen fixation was one of the many vegetable processes whose elucidation was then beyond the competence of chemistry.

Vegetation: the nuptial garland of earth and air

By the time Lavoisier published his *Elements of Chemistry*, analysis had shown that carbon, hydrogen, and oxygen were the principal constituents – Coleridge's components – of all plants, though some also contained significant amounts of nitrogen. All animal substances contained significant amounts of carbon, hydrogen, oxygen, and nitrogen. Coleridge could have come across this information while reading Erasmus Darwin's *Botanic Garden*,[101] if he had not already learned it elsewhere. Other substances were also found in vegetables, sometimes merely as traces, at other times, as in the case of silicon in the epidermis of certain reeds, grasses, and canes, in large quantities. Davy's *Elements of Agricultural Chemistry* explained that fixed alkali could be obtained from most plants; potash was "the common alkali in the vegetable kingdom," and soda occurred in some plants. Four earths, silica, alumina, lime, and magnesia, were known to occur in plants, as were the oxides of iron and manganese and compounds of sulphur and phosphorus (pp. 97, 99–101).

How did these elements come to be in the plants, subsequently to be revealed by chemical analysis? Davy believed that there was no transmutation in plant growth (pp. 154–5). Elements obtained from plants had previously been incorporated as such by plants, whether chemically or physiologically (pp. 270–4). Minerals were taken in from the soil by way of the roots, and plants gained their carbon, hydrogen, oxygen, and nitrogen "either by their leaves from the air, or by their roots from the soil" (pp. 15, 155). This was a fair general statement of the case, in line with the mainstream of chemistry since Lavoisier. But it needs elaboration, and raises problems.

First, carbon, hydrogen, and oxygen are the principal chemical components of plants. "The Plant," Coleridge wrote, "is the nuptial Garland of Earth and Air – their equation of Carbon, Oyxgen and Hydrogen . . . Now as in powers the three great Co-efficients of Nature are Gravity and Light with Warmth, as the Indifference, so in bodies, which necessarily contain each body all three, yet under the predominance of some one, Carbon most represents Gravity, Oxygen Light, and Hydrogen Warmth . . . Accordingly, in the Flower, the Crown of mature vegetative life, we have the qualitative product of Oxygen = Light in the outness and splendor of Colors, the qualit[ative] prod[uct] of Hydrogen = Warmth in the inwardness and sweetness of Fragrance."[102] Both the dynamism and the connection of chemistry with light and color, especially the association of oxygen with light, were familiar, and they were supported by recent researches. Davy's researches, for example, had shown "that the colours of flowers, depend immediately on the operation of light," and were cited approvingly by Daniel Ellis in 1811.[103]

Coleridge's writings are full of metaphors and similes based on vegetable processes. In appendix C of *The Statesman's Manual* he discusses vegetation as symbolizing the power of reason "in a lower dignity":

> I feel it alike, whether I contemplate a single tree or flower, or meditate on vegetation throughout the world, as one of the great organs of the life of nature. Lo! – with the rising sun it commences its outward life, and enters into open communion with all the elements, at once assimilating them to itself and to each other. At the same moment it strikes its roots and unfolds its leaves, absorbs and respires, steams forth its cooling vapour and finer fragrance, and breathes a repairing spirit, at once the food and tone of the atmosphere, into the atmosphere that feeds *it*. Lo! – at the touch of light how it returns an air akin to light, and yet with the same pulse effectuates its own secret growth, still contracting to fix what expanding it has refined. Lo! – how upholding the ceaseless plastic motion of the parts in the profoundest rest of the whole it becomes the visible organismus of the whole *silent* or *elementary* life of nature and, therefore, in incorporating the one extreme becomes the symbol . . . of that higher life of reason, in which the whole series (known to us in our present state of being) is perfected, in which, therefore, all the subordinate gradations recur, and are re-ordained *in more abundant honor.*[104]

This passage not only locates vegetable life within a world of correspondences, but, as Abrams has pointed out, presents an accurate and up-to-date account of the findings of experimental botanists, of the expiration of oxygen and inspiration of carbon dioxide under the influence of sunlight, of the transpiration of water vapor through the leaves and the absorption of nutrition through the roots into the rising sap.

The latter discoveries of the role of the plant's vascular system, of the flow of sap under pressure from roots to leaves, and of the emission of water vapor there, were described by Stephen Hales in his *Vegetable Staticks* (London 1727). Then came Priestley's discovery that air vitiated by animal respiration was restored by the respiration of dephlogisticated air (oxygen) by green plants. It was left to the Dutch natural philospher Jan Ingenhousz to show that light was necessary for this restoration, and "that vegetables diffuse through our atmosphere, in the sun-shine, a continual shower of this beneficial, this truly vital air." Ingenhousz also showed that only the green parts of plants accomplished this, giving off oxygen in sunlight and carbon dioxide in the dark, whereas the other parts of the plant exhaled only carbon dioxide, even in sunlight. Jean Senebier demonstrated that plants converted carbon dioxide into oxygen, and Nicolas Théodore de Saussure showed that plants obtained carbon for growth from carbonic acid (carbon dioxide). By 1813, when Davy published his *Agricultural Chemistry*, he was able to bring the results of all these experiments together: "When a growing plant, the roots of which are supplied with proper nourishment, is exposed in the presence of solar light to a given quantity of atmospherical air, containing its due proportion of carbonic acid, the carbonic acid after a certain time is destroyed, and a certain quantity of oxygen is found in its place. If new quantities of carbonic acid gas be supplied, the same result occurs; so that carbon is added to plants from the air by the process of vegetation in sunshine; and oxygen is added to the atmosphere."[105]

So far, it seems, there are no problems. Coleridge's account in his *Lay Sermons* and Davy's *Agricultural Chemistry* adequately coincide in treating of photosynthesis. And Davy, like Coleridge, accepted Stephen Hales's results, elaborating and extending them through the recent researches of Thomas Andrew Knight.[106] Carbon, hydrogen, and oxygen were straightforward in plant physiology, at least in black-box terms: One knew what when in, what came out, and what was left behind. That aspect of the external chemistry, as opposed to the internal physiological process, was known. The limitations of a chemical account of vegetable physiology were also widely recognized. Ellis, for example, having rejected the opinion "that any aeriform fluid obtains admission into the vascular system of plants by any process resembling the living function of *absorption*," found chemical explanations of the process equally useless.[107]

But if there were still some difficulties in the explanation of plant respiration, the question of elements in plants other than carbon, hydrogen, and oxygen was much more vexed. Until the source of carbon in plants was understood, it was widely held that carbon was produced by the plants themselves. De Saussure's quantitative experiments had disproved that conjecture and had stated that mineral substances in plants came through the roots from the soil. Davy and then Brande repeated these arguments: Plant growth did not involve transmutation.[108]

Coleridge was unconvinced. Why should different plants yield such different chemical analyses? Why should grasses be rich in silica? What if plants could, as had often been asserted, grow with water yet without soil?

> If an Equisetum could be raised ⟨*from seed*⟩ in flannel moistened by distilled water – or Grass or a Rattan in a pot of Soil carefully desilicated. Why should particular plants take up Sulphur, others Nitrogen, others sulphate of Lime (as clover) – many but not all marine plants common Salt – the barley nitrate of soda, the sap of the Sun flower nitrate of Potassa, Oat seeds phosphate of Lime, the Grasses & Canes Silica – while *all* have Carbon, Hydrogen and Oyxgen, live in them and reproduce, at least re-exhibit them, if it were not that *their* formative functions have an analogy with the vires formatrices of Nature, in the primary production of these *Stoffs*?[109]

These questions are very much to the point. They are among the crucial questions of vegetable chemistry, not all fully understood even today, and utterly beyond the competence of agricultural chemistry in the first half of the nineteenth century. Coleridge's questions were not only uncomfortable because pertinent yet unanswerable; they were also well informed. Davy had a chapter on mineral manures[110] that could well have provided Coleridge with much of his factual information. And before Davy had started to learn any agricultural chemistry, Coleridge was badgering Tom Poole for information that could help to make him a scientific farmer.

Carbon, hydrogen, and oxygen constituted the bulk of most plants. But there were exceptions. Davy noted that in "the reeds, the grasses, canes, and the plants having hollow stalks," the epidermis was substantial and composed mainly of silica. "This is the case in wheat, in the oat, in different species of equisetum, and, above all, in the rattan, the epidermis of which contains a sufficient quantity of flint to give light when struck by steel; or two pieces rubbed together produce sparks." Where did the silica come from? Steffens had asserted that it did not come from any lime that might have been absorbed, citing experiments by J. C. C. Schrader to support his view that elements were produced within plants.[111] Davy's experiments, carried out in 1801, were "very much opposed to the idea of the composition of the earths, by plants, from any of the elements found in the atmosphere, or in water." Coleridge wondered whether it might not be possible to reconcile these opposed conclusions by postulating that the earths in the soil might be necessary "as a sort of Ferment, or inceptive Action."[112] His interest in the problem of silica in plants is indicated by various notes, including some taken from David Brewster's paper on tabasheer, a substance found in the cavities of bamboo that consisted primarily of silica.[113]

Nitrogen fixation: animality in vegetation

Nitrogen was quantitatively a less significant ingredient in most plants than silica was; but its occurrence in all animals made its generally lesser presence in vegetables especially interesting to Coleridge. Davy had stated that plants gained nitrogen "either by their leaves from the air, or by their roots from the soil." De Saussure had suggested that nitrogen was absorbed through the roots, a position shared by Ellis, who found no absorption of atmospheric nitrogen by growing plants. This was unfortunate, because other had found no absorption from the soil. "So that while Mr. Ellis bars up one door (the pulmonary) and derives all the constituents of the plant from the Soil, Messrs. Brande and Brodie ~~double lock~~ stultify the other inlet of conjecture. For here no [nitrogen] exists – ergo, none [can] enter – and yet there the mysterious Elements *is*." If the experimentalists were all correct in their negative findings, nitrogen had to be produced by some action within the plant. Chemistry was totally unable to account for nitrogen fixation, and Coleridge was acute in seizing on this weakness.[114]

Nitrogen in plants also fascinated Coleridge because it was "an alien, an antedated Animal," in the vegetable kingdom. It was a principal constitutent in "the gluten of Wheat, which is almost an artefact of man by animal manures." He took notes on gluten from Brande, who called it vegetable albumen. Coleridge found gluten fascinating because of its high nitrogen content, suggestive of animality in vegetation; the presence of gluten in both bread and wine also provided him with a resonant symbol of body and spirit in transubstantiation. He manifested a similar interest in mushrooms which were rich in nitrogen, were plants but never grass green, and, as a bonus in paradox, were cryptogams, plants with hidden sexual organs whose mode of reproduction was puzzling, and which therefore resisted convincing integration into Linnaeus's scheme of sexual classification.[115]

The problem of the source of nitrogen in plants was matched by a complementary problem in animals, one troublesome to Coleridge but remarked by few if any others. "One most pregnant fact," he observed, "that [there is] an equal quantity of Nitrogene in the chyle of graminivorous animals as of carnivorous – & it has been proved that no nitrogene passes into the system thro' the Lungs."[116] Chyle was produced in animals by a sequence of steps in which food entered the stomach, was converted to chyme, and passed thence to the small intestine, where it was blended with bile and separated into two portions, one being excreted via the large intestine, the other being "absorbed by the lacteals, which terminate in the common trunk, called the *thoracic duct*." The latter portion was the chyle, which, mixed with lymph in the thoracic duct, "poured into the venous system."[117]

The food of graminivorous animals contained little nitrogen; their chyle was rich in nitrogen, which did not enter the body through respiration. Once again Coleridge raised the possibility of the production of nitrogen in living beings. This was one source of his interest in the chemistry of digestion. He read Brande's account, just given, and also Everard Home's "Observations on the structure of the stomachs of different animals, with a view to elucidate the process of converting animal and vegetable substances into chyle," and probably also his "Hints on the subject of animal secretions." These papers clarified contemporay physiological theories for Coleridge, and, with Brande's work, furnished him with the chemical analyses that were the basis of his queries.[118]

Now chyle seemed very similar to blood. Brande noted that "it is deficient only in colouring matter, and the albumen which it contains differs a little from that existing in the blood itself." Hatchett, whose analyses much impressed Coleridge, had tried to show that albumen was the "original animal substance." And albumen was rich in nitrogen. It is therefore not surprising to find Coleridge interested in albumen, "the firmamental principle of the animal," rich in carbon as well as nitrogen, and in the coloring matter of the blood. Lymph, blood, albumen, cerebral substance, gelatin – all these could be scrutinized for their chemical constitution, but chemistry gave little insight here, being at best suggestive.[119]

Chemistry and the ascent of life

When Coleridge came to consider Hatchett's analyses of shell and bone, he was able to transcend the limits of chemistry. Shells were mostly carbonate of lime, with very little animal matter, "but that (n.b.) gelatine," which he thought might be "potentziate Albumen." Then came mother of pearl, with sixty-six parts carbonate of lime, and the rest animal matter that was all albumen. Higher in the scale of life, hen's eggshell and lobster claws contained mostly carbonate but also phosphate and animal matter. The same chemical progression, from gelatin to albumen and from carbonate to phosphate, was repeated in the zoophytes. And bone contained much more phosphate than carbonate. "A beautiful Harmony – and all ascending," exclaimed Coleridge. The proportions of gelatin to albumen, and of phosphate to carbonate, increased as one came to higher and higher organisms.[120] Coleridge's concept of a scale of living beings was founded upon his view of the development of the power of life. Chemistry, pursued through geology "in the organismus of our Planet," led Coleridge into the lives of individual organisms, where it was subordinated to a higher triad of powers, reproduction, irritability, and sensibility.

7

LIFE

CROWN AND CULMINATION

Coleridge and the doctors

Coleridge's cosmos was a living one, informed by the power of life, and reaching its climax in man at the summit of the terrestrial creation. Coleridge's concern with the ascent of life complemented his probing into his own human nature and his fascination with the relations between mind and body. These interests drew him to medical works, and accidents of personality and health brought him into extended intercourse with medical men. Many of the sixty or more physicians and surgeons among his acquaintance were interested in powers, in the theory of life, in zoomagnetism, and in psychosomatic diseases.

Coleridge's brother Luke (1765–90) had been a surgeon in the London Hospital. Coleridge recalled his desire to emulate his brother: "Every Saturday I could make or obtain leave, to the London Hospital trudged I. O bliss, if I was permitted to hold the plasters or to attend the dressings, I became wild to be apprenticed to a surgeon. English, Latin, yea Greek books on medicine I read incessantly. Blanchard's Latin Medical Dictionary I nearly had by heart." When in 1797 he stated the preparation needed for writing an epic poem, he included the acquisition of a thorough knowledge of anatomy and medicine.[1]

His studies in Germany in 1799 took him to Blumenbach's lectures on physiology, anatomy, and natural history. When he returned to Bristol, it was to find Beddoes and Davy busy with pneumatic medicine, and with access to the patients in the Bristol Infirmary. He had his children vaccinated against smallpox, solicited material for an article from Jenner, and maintained an interest in medical matters.[2]

In 1816 he had the good fortune to move in with James Gillman, and for the rest of his life was in almost daily intercourse with medical men. He praised Gillman's work on hydrophobia, composed the *Theory of Life* and the *Essay on Scrofula*, met Abernethy, and became involved in his continuing fight with Lawrence.[3] He read widely in medical texts and journals, displaying an interest in medical topics from quarantine to cancer. Under the tutelage first of Gillman and then of Joseph Henry Green he was enabled to make trenchant and informed comment on contemporary medical debate. The transaction was not one-sided. Green's works bear the imprint of Coleridge's thought. Gillman likewise sought to apply dynamic philosophy to medicine – for example, elucidating the nature of inflammation by using the logic of trichotomy.[4]

The incursion of powers into medicine was complex, through Schelling via Haller, John Brown, C. F. Kielmeyer, and others. Here was one of two convergent routes that enabled Coleridge to conceive the life sciences dynamically, uniting physiology and natural history.

From irritability to dynamic physiology

When muscular tissue is stimulated externally, it contracts. Albrecht von Haller (1708–77), poet and physiologist, attributed this contraction to irritability.[5] He also explored the relation between nerves and muscles, finding that nerves could respond to stimuli arising from muscular action through a force or property that he called sensibility.

Haller's work led to a spate of researches on the nervous system. William Cullen (1712–90) in Edinburgh studied the nerves as regulators of physical functioning. John Brown (1735–88) was among Cullen's pupils, and was accused of stealing his ideas. Brown's doctrines, published in his *Elements of Medicine*, were known as the Brunonian system, and were republished as such in 1795, with a biographical preface by Thomas Beddoes. Brown postulated that every living being was endowed with its portion of excitability, the principle on which life depended. Excitability, located in the medullary portion of the nerves, had to be maintained in proportion to the stimuli acting on the system, or illness, even death, would ensue. Thus the treatment of disease that he advocated was directed toward adjusting stimuli so as to restore its proper balance with excitability in the patient, whether by augmenting or diminishing it.[6] Thomas Beddoes considered Brown's system far from perfect, but he may have been partly influenced by it in prescribing opium as a stimulant. Erasmus Darwin adopted Brown's theories in his *Zoonomia* (London, 1794). Coleridge was a victim of Brunonian physic.

Brown's theories were widely known in Germany by the time Coleridge reached Göttingen. He read a long review of works on the Brunonian System, spread over twelve numbers of the *Allgemeine Literatur-Zeitung* (Jena and Leipzig, February 11–20, 1799). He had probably been stimulated to read the article by Beddoes's critical interest in Brown's theory. The subject was clearly a controversial one in Göttingen, for in 1802 the cavalry was called out to put down rioting between Brunonians and their critics.[7] Physicians were not the only ones to adopt the system: Philosophers did so too. Schelling in 1800 went to the clinic in Bamberg in order to study the methods inspired by Brown's theories. He also knew Carl Friedrich Kielmeyer's lecture of 1793 on the interrelation of organic forces, in which the threefold division of organic forces into reproduction, irritability, and sensibility was advanced. Schelling appears to have modified Brown's scheme and fused it with Kielmeyer's, dividing excitability into active and passive components, receptivity and activity. He further divided receptivity into sensibility and irritability, the latter externalized as reproduction. And he varied his formulation by sometimes calling the power of reproduction *Bildungstrieb*, thereby incorporating Blumenbach's theory of generation, also derived from Haller, first published in 1791, and advanced in the lectures on physiology that Coleridge attended.[8]

In adopting the threefold classification of reproduction, irritability, and sensibility, Schelling was asserting a hierarchy of organic powers congruent with his doctrine of three dimensions, length, breadth, and depth, and of three physical powers, magnetism, electricity, and chemical process. Correspondences between the different realms of nature were complete. When Coleridge encountered these correspondences, he would find them all the more convincing because they incorporated the medical theories of Blumenbach, Darwin, and, with some qualifications, Beddoes.

Coleridge found Brown's theory fundamentally unsound in its view of life as a *forced* state of excitability. Life, whatever it might be, was natural and unforced. But he accepted Schelling's view of the three forms of the vital principle, representing sensibility as the product of reproduction and irritability, just as galvanism was the product of electricity and magnetism.[9]

He was repeatedly unhappy with Schelling's terms. He wondered whether it might not be better to substitute centrality for sensibility, the highest form of the power of life, for this would at least indicate its mode of production as the central power at its level in the compass of nature. Again, reproduction was common to all animals and vegetables.[10] Irritability, with few exceptions, was lacking in vegetables but present in even the lower animals and insects.[11] Men and the higher animals exhibited sensibility. The three terms might therefore be rendered as vegetivity, instinctivity – because even bees and ants possessed instinct – and centrality or "the idiozoic" – because men existed "*for* themselves." And elsewhere, he gave it as proven that "Vegetivity = Zoo – Magnetism, Insectivity = Zoo – Electricism, & Animality = Zoo – Galvanism." Coleridge saw the ascent of powers, from reproduction to sensibility, as the ascent of life. Insects, for example, pass "from the highest forms of Veg. Repr. into Irritability, & present Irritability tending to sensibility."[12]

Such a hierarchical conception of organisms has much in common with the schemes of the *Naturphilosophen*, especially Steffens and Oken. Steffens was guided primarily by the notion of an ascent of power in constructing his scheme of living beings. Oken adopted a classification in which organisms were ranked according to their principal organs or anatomical systems. This led to a hierarchical ranking of classes that exhibited internal parallels with one another. The principal organs were skin, tongue, nose, ear, eyes; the corresponding senses were touch, taste, respiration and smell, hearing, and sight; and the corresponding classes were invertebrates, fishes, reptiles, birds, and mammals.[13]

In his *Lehrbuch der Naturgeschichte*, Oken arranged species and genera according to these principles of classification. And this led to an arrangement in which animals "become nobler in rank, the greater the number of the organs which are collectively liberated or severed from" the animal kingdom conceived as one animal.[14] This presented an ascent of life having something in common with Coleridge's arrangement based upon the principle of individuation, and he therefore sought to adopt what was valuable in it, while rejecting what he deemed fanciful fictions.

Coleridge welcomed the correlation of ascent in the hierarchy of life with the development of organs and their corresponding systems. In his 1815 scheme of the compass of nature, he postulated an organic level, in which glandular and venous, arterial and muscular, and nervous systems corresponded to reproduction, irritability, and sensibility. And, in decidedly exploratory fashion, he attempted to extend and elaborate these fundamental correspondences, taking into account the physiological functions of the different systems. He considered, for example, "Compound Systems divided into three according to the great Cavities," which he identified as abdominal, pectoral, and cerebral. The abdominal system was "formed by the LIVER, Spleen, Pancreas, Stomach, Gut." Its functions were "Digestion, Assimilation, Secretion, Excretion." Veins were preponderant, and the common object of the whole system was "Reproduction."[15]

All this, however, was frankly exploratory. He was looking for trinitarian arrangements of powers, systems, organs, and processes, all correlated with a hierarchical classification of organic beings, agreeing with the facts of physiology and anatomy, and conforming to the principles of dynamic logic.[16]

Organization

In marginal notes to Boehme and Boerhaave, Coleridge referred to the debate between followers of Blumenbach and Hunter about the relation among bodily fluids, structure, and life. Hunter had asserted that life was in fluid blood, from which organization arose. Blumenbach also held that blood was "the chief and primary fluid," and that solids were derived from fluids. He saw however that solids and fluids were interdependent, and that in living bodies the fluids were organized in some kind of structure. Coleridge thought that the whole argument was based on the misuse of words, and that "nothing wholly fluid or wholly solid does or can live." The question of the relations between bodily fluids, and of fluids to solids, was nevertheless a crucial one for contemporary physiology.[17] Cuvier expressed the problem and its importance succinctly in his *Lectures on Comparative Anatomy* (trans. W. Ross, 2 vols., London, 1802, *1*, 33): "In a word, all the animal functions appear to reduce themselves to the transformation of fluids. In the manner in which these transformations are produced, the real secret of the admirable oeconomy of animals consists, as health depends upon their perfection and regularity."[18] The relations of life to fluids and of organism to organization were complex; Coleridge had to try to elucidate them in forming his theory of life.

Schelling's concept of the cosmos as an animated organism furnished the central metaphor of *Naturphilosophie*. The organism, prior to and opposed to mechanism, was at once unified and total, having all its parts subordinated to the whole. Within it, organized bodies and organization arose from the limitation of productivity; Schelling considered that organization was "nothing more than the *arrested* stream of cause and effect."[19] Individual organized bodies were thus products of the striving of the spirit or power of the universal organism. The *Naturphilosophen* adopted this broad formulation, elaborating it in different ways. Steffens's view of the animal creation as a hierarchy of organized beings, ranked according to the degree of development within them of the power of individuation, was compatible with Schelling's doctrine.[20] So was Oken's view of the animal creation as a hierarchy of organisms constituted from aggregations of vesicles of primal slime; as one ascended the scale, one found more elaborate organs more fully subordinated to the purposes of the organisms of which they were parts.

Coleridge's *Theory of Life* drew on Oken, Steffens, and Schelling, who in turn were indebted to Kant. In the *Critique of Teleological Judgment*, Kant had explained that in organized beings, "every part is thought as *owing* its presence to the *agency* of all the remaining parts, and also as existing *for the sake of the others* and of the whole, that is as an instrument, or organ." He was echoing what Coleridge called the "Aristotelian axiom . . . that the whole is . . . prior to its parts." Kant added that each part "must be an organ *producing* the other parts – each, consequently, reciprocally producing the others." Here were riches for subsequent dynamic philosophy. Coleridge, even more than the *Naturphilosophen*, must have welcomed Kant's repeated assertion that reason led us to see organized beings as supplying natural science "with the basis for a teleology." The existence of organisms, inexplicable by "the mere mechanism of nature," implied "causality according to ends," "a cause acting by design," and "a creative understanding."[21]

Kant also paid tribute to Blumenbach's contributions to epigenesis through the concept of a formative impulse in living creatures. Blumenbach returned the compliment by adopting Kant's defintion of organized beings. Cuvier adopted it, too. Once comparative anatomists and physiologists embraced Aristotelian and Kantian notions, allowing the possibility that there might be some kind of vital force or impulse, it became necessary to tackle the question of the relation between organization and life. Cuvier had no doubts that the mechanical part of organization was "only the passive instrument of vitality."[22]

John Hunter thought so, too. "I shall endeavour to show," he wrote, "that organization, and life, do not depend in the least on each other; that organization may arise out of living parts, and produce action, but that life can never rise out of, or depend on, organization . . . mere organization can do nothing, even in mechanics, it must still have something corresponding to a living principle; namely, some power." Richard Saumarez, to whom Coleridge referred enthusiastically in the *Biographia*, followed Hunter. So too, in the main, did John Abernethy.[23]

The question of organization was central to the Lawrence–Abernethy debate, arising partly from Lawrence's determination to keep physiology empirical, even positivist.[24] The *Quarterly Review* (*22* [1819], 3) claimed that Lawrence said that vital properties were derived from organic structure, and that differences of structure constituted the only difference in the faculties and powers of different organisms. But as his article on life in Rees's *Cyclopaedia* (*20* [1819]), makes clear, his position was more complex. "The properties of any living organ are of two kinds: the one immediately connected with life, beginning and ending with it, or rather forming its principle and essence; the other connected to it only indirectly, and appearing rather to depend on the organisation, on the texture of the part." Sensibility was a vital property, whereas contractility arose from organization.

Lawrence's more extreme statements in his lectures at the Royal College of Surgeons must therefore be seen as part of a polemic with Abernethy. Coleridge's entry into the debate in the first part of the *Theory of Life* must also be seen in this context. His concern with establishing the subordination of organization to life was also a concern with exposing the inadequacy of positivism and empiricism in science.

These were his more distant goals in composing the *Theory of Life*. In 1816 he wrote to Gillman that his object was "to connect Physics with Physiology, the connection of matter with organization, and of organization with Life." Similarly, in the *Opus Maximum* he asked "whether in life we may not find the conditions of universal organization and again in universal organization the conditions and the solution of mechanism."[25] There is constancy and continuity of method in Coleridge's inquiry into life and organization from 1816 until the late 1820s. He viewed organization in Kantian and Aristotelian fashion as an interdependence of parts, all of which were means to an end, and which in their unity also served an end. He therefore followed Hunter, Cuvier, and Abernethy in arguing against the possibility that organization could produce life.[26]

Organization implied purpose, whether it was the all-embracing purpose directing the "organific activity of nature," or the purpose associated with the individual organism and "the activity of organization ab intra." The distinction between the activity of nature and the growth of individual organisms made clear Coleridge's belief that organization could exist outside living organisms; so, too, did Gillman's statement, accepted by Coleridge, that organization was potential life, "Life *Actual*" having "no organ." The concept of organization, interpreted through dynamic logic, bore directly on the *Theory of Life*.[27]

Life actual

What was life? The question had been an urgent one for Coleridge from the very start of his interest in science. On New Year's Eve, 1796, he wrote to his radical friend John Thelwall, who had lately written "an Essay on Animal Vitality": "Dr. Beddoes, & Dr Darwin think that *Life* is utterly inexplicable, writing as Materialists – You, I understand, have adopted the idea that it is the result of organized matter acted on by external Stimuli. – As likely as any other system; but you *assume* the thing to be proved – the '*capability* of being stimulated into sensation' *as* a *property* of organized matter – now 'the Capab.' &c is *my* definition of *animal Life*." Then, after dismissing the opinions of several authorities, Coleridge admitted, "*I, tho' last not least, I* do not know what to think about it – on the whole, I have rather made up my mind that I am a mere *apparition* – a naked Spirit! – And that Life is I myself I! which is a mighty clear account of it."[28] The problem, insoluble within the confines of science, continued to fascinate Coleridge. He read H. S. Reimarus's account of the severed head of a rattlesnake that still sought to strike and bite, and Pierre Lyonnet's observations of a wasp whose head and hind half continued to eat and sting after they had been cut asunder. "How," Coleridge wondered in 1804, "shall we think of this compatible with the *monad* Soul? . . . Or is there one Breeze of Life – at once the soul of each & God of all?"[29]

Over the next twelve years, he moved toward Hunter and away from the pantheism hazarded and rejected in "The Eolian Harp." In 1820, he was elaborating the implications of vital dynamism, distinguishing te life of organisms from the life of nature. Both had their own entelechy, but in the former it acted determinately, and in the latter indeterminately. All this, Coleridge concluded, was clear – "clear therefore that however endless the grades of descent may be in living things, there must ~~be~~ still be a chasm between the lowest, & the inanimate."[30]

The relation between organized inanimate and animated nature was nevertheless troublesome. Clearly a crystal was not alive in the same way as an animal; yet it was organized. Gillman had not yet offered his useful definition of organization as potential life,[31] and in the *Theory of Life*, Coleridge had stressed the continuity of life throughout nature, in its ascent from mineral to man. But still a distinction had to be made between the potential life in a crystal and the actual life of animals and plants. What was the nature of the chasm between the lowest living organism and inanimate matter?

One promising approach to this nagging question was adumbrated in a note made in or around 1820: "*Law.* That Nature leaves nothing behind. – The exceptions apparent only or such as confirm the Rule. Memo. Apply this law to Chrystallisation ÷ Life." This cryptic entry is rendered intelligible by his explanation that the sign ÷ "means a division which is yet not an absolute chasm. It ~~flows~~ dips underground, and then filtrates for ascension, and re-appears as a spring in a diverse *Kind.*"[32] The ascent of the power of life was, then, only apparently interrupted. It manifested itself in its lowest form in crystals, then sank back, reemerging in altered kind in the lowest living organisms. Distinction, difference, and unity were thus maintained in a pattern that was repeatedly exhibited in the *Theory of Life*, the power of life dipping down after one class of organisms, to reappear, transformed, in the lowest members of the next distinct class. The overall ascent of life thus exhibited apparent discontinuities masking real continuity. Hence there was an essential diversity in kind between crystals and protozoa – yet only a narrow gap between them in the ascent of the power of life. This resolution of the problem enabled Coleridge to build upon without invalidating the *Theory of Life*.[33]

Life as a power: problems of classification

Blumenbach had presented his view of life as a power in the course of opposing the doctrine of preformation, whose most extreme version was the theory of *emboîtement*; according to this theory the generations of mankind were all contained in Eve's ovaries, one within the other like Chinese boxes.[34] Blumenbach thought that this was absurd and advocated instead epigenesis, the progressive formation within the mother or egg of each new generation. Generation, nutrition, and reproduction were all directed by a *Bildungstrieb*, *nisus formativus*, or formative impulse, whose cause, like that of Newton's gravitation, was hidden, but whose effects were manifest. Coleridge absorbed the lesson and, with his perpetual search for correspondences between powers of mind and powers in nature, asked about the connection of the imagination with the *Bildungstrieb*, a power that strove in all living things.[35]

Hunter saw life itself as a power working in all living things. He was in fact not nearly as explicit as Coleridge made him. But Coleridge would have him so, deeming him "the profoundest, we had almost said the only, physiological philosopher of the latter half of the preceding century." Hunter "did not make an hypostasis of the principle[s] of life, as a . . . phaenomenon . . . ; but . . . he philosophized in the spirit of the *purest* Newtonians, who in like manner refused to hypostasize the law of gravitation into an ether . . . The Hunterian position is a genuine philosophic IDEA." Hunter was a true genius, whose idea of life had the simplicity of genius, and made him the morning star of "the Constellation of Great Minds, who . . . have founded the Science of Comparative Anatomy & Physiology."[36]

The most eloquent statement of Hunter's view of physiology and comparative anatomy was in the selection and arrangement of specimens for his museum, rather than in his writings – hence Coleridge's success in attributing to Hunter ideas clearly beyond his utterance. But there was a theory in Hunter's arrangement. It was left to Everard Home to interpret it, in his *Lectures on Comparative Anatomy; in Which Are Explained the Preparations in the Hunterian Collection, Illustrated by Engravings* (6 vols., London, 1814–28).[37] He explained that

> this arrangement . . . is peculiarly adapted for those who mean to prosecute comparative anatomy. It is an attempt to class animals according to their vital and other internal organs, forming the whole into one regular series of gradations, beginning with those animals that have the most simple structure and tracing them upwards as their parts become more numerous till we arrive at man, the most complex in his organs, who forms the highest link of the chain.
>
> According to this plan, Mr. Hunter's system begins with animals that have nothing analogous to a circulation; then follow others which have some approach towards one; and afterwards animals in which it is distinct; and so on through all the complications which lead by almost imperceptible steps to man, in whom the heart is the most compounded.
>
> All the organs of an animal body are arranged in distinct series, beginning with the most simple state in which each organ is met with in nature, and following it through all the variations in which it appears in more complex animals. [*1*, 6–7]

Here, as in Oken's system, comparative anatomy led to a classification embodying a theory of the ascent of life.

All the authors whose accounts of living nature Coleridge favored shared Hunter's view that comparative anatomy was the key to zoological classification. They also, like Hunter, recognized the inapplicability to nature of classifications embodying a single and continuous scale of ascent. Blumenbach complained of the Linnaean system that it often disregarded anatomical structure, "which should form the basis of a natural classification." And in his *Manual of the Elements of Natural History*, which Coleridge had once thought of translating, he remarked that although the "common metaphor" of a scale of beings might aid the formation of a natural system of classification, it had not formed part of "the plan of the Creation." Species in which the sexes differed widely, and such problematic animals as the tortoise, simply did not fit any such scale. Coleridge, following Blumenbach and Cuvier, observed in a marginal note in another of Blumenbach's works:

> The fault common to the systems & systematizers of Natural Hystery is . . . not so much the falsehood nor even the unfitness of the guiding principle, diagnostic or teleological, adopted in each; as that each is taking [*sic*] as the only one, to the exclusion of others, and as adequate per se to exhibit the whole of the Subject – the Systematizer forgetting that Nature may pursue a hundred Objects at the same time, and each by a different Line or Chain of Facts from the first Hint of the Purpose to be realized to the Structure or Organ in its most perfect form, & in which the final cause of the whole series . . . has attained its' full evidence, and determines the place of all the intermediate Facts.[38]

The systems that Coleridge condemned stood in contrast to Cuvier's. Cuvier sought to classify living beings according to the totality of their structure, employing the principle of the subordination of characters, taking account of the correspondence between exterior and interior forms, and generally seeking to base his classification on the "physiological integration of function rather than an abstract morphological plan of structure."[39] He welcomed the increasing interdependence of natural history and anatomy, and used the examination of organs, function, and structure to establish a natural classification. Different organs did not all follow the same order of gradation; so he decided that there were as many series as there were regulating organs. This was the origin of his four *embranchements*, vertebrates, molluscs, articulates, and radiates.[40]

Oken's scheme in his *Lehrbuch der Naturgeschichte* was similarly based upon organs, and postulated five principal groups instead of Cuvier's four. Man appeared as the crown of nature's development, comprehending everything that had preceded him and representing the whole world in miniature. The plan of nature's ascent thus achieved a total unity, all lower organisms exhibiting part of the pattern of man. Unity was also a consequence of Oken's view of the whole animal kingdom as one great animal, and of his idea that all organic beings were built of protoplasmic cells. Thus, although Oken's scheme was more complicated than Cuvier's, it possessed a greater conceptual unity. Oken criticized Cuvier's system for its "horrible lack of symmetry." Cuvier's scheme was essentially anatomical, Oken's was both anatomical and physiophilosophical.[41] Coleridge commented on the debate: "Inattention to the impossibility of graduating per ascensum the classes, orders and genera of Organized Individua under any one form of Abstraction, or for any single point of view, is the true cause of the controversy between Oken . . . with older Naturalists on his side, and M[e]ckel, Spix, Sweigger, & Goldfus[s], the followers on this point of the French Naturalists, Cuvier, Lamarck, Dumeril & Blainville –. Doubtless Oken is in the right – in the main, because he uses more the *unseen* Eye."[42]

Coleridge, in classifying naturalists, was clearly a lumper rather than a splitter. But he was right to see Cuvier as directed more by observation than Oken, with his notions of the great animal, protoplasmic slime, and man the pattern of all living nature. Coleridge rejected Oken's fictions, while being attracted by his view of the ascent of living things and of man as the crown of creation.[43] And Oken's magisterial handbook was full of that accurate, detailed information about nature that Coleridge so relished. But it was Steffens, with his classification through an ascent of powers, who most captured Coleridge's assent. Steffens developed the doctrine of the three powers of reproduction, irritability, and sensibility into an all-embracing zoological classification and philosophy, from which Coleridge drew freely in composing his *Theory of Life*.

Coleridge's goal was succinctly expressed in a note of 1820: "Nature-history on a new scheme of Classification – first, of Powers, each Power subdi~~vided~~stinct into its Force & ~~all~~ each Force again according to the *proportion* of its predominance in the Species of animals subsumed; & then according ⟨to⟩ the kind (N. or W.) and degree of its modification . . . then the Divis. & Subdivis . . . according to the circumstances or habitats." The powers were the familiar three; habitats were water, land, and air, including combinations among them, for example, "Water + superaqueous Air (Whales &c.)" "This *Combination*," Coleridge explained, "is most important . . . in reference to the . . . Principle of the Powers – thus Fish pass from a lower form of Reprod. (Water plants) into Sensibility – Insects from the highest forms of Veg. Repr. into Irritability, & present Irritability tending to Sensibility – while the Bird in which the sensibility of the Fish exists is promoted, pauses, as it were, to combine itself with the irritability of the Insects & their Art-instincts and then commences the harmonising & equilibrising of the three powers in the Udder of the Mammalia."[44]

Natural history: instinct, teleology, and types

Coleridge did not limit his reading in natural history to works of classification, anatomy, physiology, and philosophy. He read with delight and fascination Gilbert White's *Natural History and Antiquities of Sel-*

borne, William Kirby and William Spence's *Introduction to Entomology* (1815–26), François Huber's *Natural History of Ants*, works by Lyonnet, Jan Swammerdam, Reimarus, and many others.[45] This was in part to feed his constant appetite and relish for the infinite variety of living things. But he had also a clear perception of the ideal behind the real, an abiding philosophy that gave widening significance and resonance to the least detail of fact in nature. When Coleridge went to the ant, it was not only with a naturalist's eye, but also with a view to extending his comprehension of the role of powers in mind and nature, the correspondences among them, the teleology directing them, and the language embracing and incorporating them.

Coleridge pondered the nature of instinct in ants, associating it with irritability, and thus placing it in correspondence with the power of electricity. But there was also another dimension to instinct. The entire social organization of bees and ants and the choice of a particular plant for a butterfly to lay its eggs on were examples of instinct at work, exhibiting a rudimentary form of intelligence. Instinct, he suggested, might be the light of intelligence acting in a creature, "but not for it, as an Object distinguishable therefrom[;] when the Light is one and the same with the *Life*, we say that such a creature is endued with *Instinct* – Light indistinguishable from Life is Instinct, not Reason." Instinct was akin to understanding, as opposed to reason. The ascent of life became, through this perspective, an ascent of powers of mind, and the ant offered clues to this process.[46]

Once mind was admitted as nature's informing guide, natural history acquired added significance. "A campaign of the Duke of W[ellington] consisting wholly of the visual & auditual phaenomena? So a Nat. History."[47] It was not enough merely to describe things and events; one needed also to grasp the strategy guiding them, their function, and their goal. The works of nature, like those of man, were subject to the operation of final causes. Coleridge admitted that it would be premature for the naturalist to substitute final for efficient causes; yet even the enlightened naturalist admitted "a teleological ground in physics and physiology."[48] Kant's doctrine, that every organism could be understood only teleologically, was one to which Coleridge fully subscribed. Every created being developed according to a plan. "The Larva of the Stag-beetle," Coleridge wrote to his nephew Edward in 1826,

> lies in it's Chrysalis like an infant in the Coffin of an Adult, having left an empty space half the length, it occupies – and this space is the exact length of the Horn that distinguishes the perfect animal, but which, when it constructed it's temporary Sarcophagus, was not yet in exis-

> tence. Do not the Eyes, Ears, Lungs of the unborn Babe give notice and furnish proof of a transuterine, visible, audible, atmospheric world? . . . But likewise – alas for the Man, for whom the one has not the same evidence of Fact as the other! – the Creator has given us spiritual Senses and Sense-organs – Ideas, I mean! . . . and must not these too infer the existence of a World correspondent to them?[49]

Everything living in nature, material and spiritual, was purposive.

There was in Coleridge's day a widely prevalent belief that the appearance of design in nature could lead men from nature to an understanding of the goodness, power, and wisdom of the creator – from nature to nature's god. This natural theology, popular through the late seventeenth and eighteenth centuries, expressed succinctly in Paley's *Natural Theology* and more elaborately through the successive volumes of the Bridgewater *Treatises,* certainly rested upon teleology. But it was logically and theologically offensive to Coleridge, who complained of its gross sophistry, its "degrading conceptions of the divine Wisdom," and its inutility. He mocked it for the limitations it imposed on human inquiry, and for its attempt to deduce the existence of god a posteriori. And he complained that "to deduce a Deity wholly from Nature is in the result to substitute an Apotheosis of Nature for Deity." The contemplation of nature required greater reach of mind and greater humility than the natural theologians allowed or evinced.[50]

Coleridge repeatedly remarked nature's self-anticipation – instinct in ants foreshadowing understanding in man, the anatomy of the apes foreshadowing human anatomy, higher vegetables foreshadowing lower animals. Here were evidences of guiding purpose. But between the lower anticipation and the higher manifestation there was always an absolute separation. "As the analogy which cannot be denied and as the chasm which in spite of the undeniable Analogy cannot be filled up or bridged over, between the granulation of Metals . . . and the formation of the Infusoria . . . ; so neither can the Analogy or the impassable Chasm between the highest Orders of Animals and the Man. They are *Types* not Symbols, dim Prophecies not incipient Fulfilments."[51] Coleridge was here using "type" in both its theological and zoological senses. The former held types to be prefigurations or representations in prophetic similitude. In the latter sense, the concept of type was a central organizing idea in zoology in the first half of the nineteenth century;[52] or rather, it furnished several different organizing ideas. A single member of a group could serve as the type, model, or representative specimen of that group, simplifying descriptive classification in natural history. A group of organisms could be distinguished by sharing a common structure or character, which became the type for the group; Cuvier, for example, described four morphological types of nervous system as the basis of his four *embranchements*. And not only the characteristic structure, but also the group characterized, could be regarded as a type. There was also a more restricted use of the concept of type, in which a particular specimen in a collection of specimens served as the type, model, and name carrier of a species.

These were the principal uses of the concept of type among natural historians in Coleridge's day. His own explicit dissociation of type from symbol makes it clear that his use of type is closest to Cuvier's common significant form. Cuvier's interest in morphological types was primarily in their function; others, like Geoffroy Saint-Hilaire, were more concerned with form. *Naturphilosophen* like Oken and Carus sought an a priori explanation of the community of forms in living organisms. Coleridge, though sharing their emphasis upon form and their desire to integrate types into a rational plan embracing all nature, went to an older tradition of idealism in developing his views. He went to the Neoplatonism of Porphyry, Plotinus, and Iamblichus, writing of "significant Forms . . . or organiz~~ed~~ation, ~~Bodies~~, contemplated as so many Types or Characters impressed on animal bodies, or in which they are, as it were, *cast*. Now Types and Characters variously yet significantly combined form a Language. – The Types of Nature are a Natural *Language* – but a Language is no more conceivable without reference to Intelligence than a ~~system~~ series of determinate actions without reference to a Will, if not immediately yet ultimately. An intelligible, consistent and connected Language no less supposes intelligence for its existence than it requires it for its actual intelligibility."[53] For Coleridge, the contemplation of animated nature through the idea of types revealed the ascent of powers, purpose, activity of mind and will, and something of the very language of creation, the power of the word, whereby the worlds of nature, mind, and creation coalesced.

The *Theory of Life*

When Coleridge wrote the first draft of the *Theory of Life*, his immediate concerns were the Lawrence–Abernethy debate and the application of the philosophy of Schelling and Steffens to the hierarchy of living beings. He had already confronted the problems of the ascent of powers and the role of mind in nature. He had learned about contemporary investigations into the relations between medicine and physiology, physiology and anatomy, and anatomy and natural history. At the same time, he had formulated his view of the interdependence of science and philosophy and of the subordination of both to theology. Thus, as early as 1816, the *Theory of Life* appears in conception as an integral part of a lifelong intellectual endeavor, a stage in Coleridge's preparation for his great synthetic work, of which surviving manuscript volumes of the *Opus Maximum* and "On the Divine Ideas" form a substantial part.[54]

Coleridge embarked on the *Theory of Life* in 1816, at the start of approximately a decade of scientific reading and metascientific thought. That decade saw a massive although selective enlargement of his knowledge of science, together with a progressive disenchantment with *Naturphilosophie* and a corresponding increase in the trinitarian form and content of his thought. Coleridge could more confidently have written his essay on life in 1826 than in 1816. But there is a sense in which the essay, significantly unpublished in his lifetime, was a program for ten years of reading and thinking. He returned successively to the problems and sciences contained and referred to in the *Theory* of 1816, moving from the construction of matter, physics, and cosmogony, through the earth sciences and chemistry, to the sciences that studied animated nature. Finally he came to man, whom he, like Oken on the one hand and Christian theologians on the other, saw as the culmination and crown of organized creation.

The bulk of his scientific reading can be seen as at once development and recapitulation of arguments in the *Theory of Life*.

"The most difficult problem," Coleridge observed in 1827, "is that of the Bearings of Generic Nature on Germinal Power as the Principle of Individuality."[55] This was the problem he addressed in his modestly entitled *Hints towards the Formation of a more comprehensive Theory of Life*. Having disposed of Bichat's and Lawrence's formulations, he proposed to explain life as a power or tendency, through the presentation of "an ascending series of corresponding phenomena as involved *in*, proceeding *from*, and so far therefore explained *by*, the supposition of its progressive intensity and of the gradual enlargement of its sphere, the necessity of which again must be contained in the idea of the tendency itself."[56]

He set about this enterprise in the most comprehensive fashion. "What is Life? Were such a question proposed, we should be tempted to answer, what is *not* Life that really *is*?" (p. 38). Here was a radical alternative to the constricted answers posed by mechanistic physiology. Coleridge saw an ascent of life from the metals, whose galvanic properties he associated with irritability, all the way up to man with his excitability. In adopting this approach, he stressed the unity and continuity of ascent in a way that he would later see as blurring the distinction between the life of nature, active throughout the terrestrial organism and the entire cosmos, and the lives of individual organized beings. But in 1816, reacting against Lawrence, he minimized what he later stressed as a complete *saltus* between chemistry and physiology. As for the manner of the ascent of life, it was one with the ascent of powers in Coleridge's hierarchical classifications of the sciences (p. 41).

The most comprehensive formula that he could give for life in its ascent was "the *power* which discloses itself from within as a principle of unity in the *many*" (p. 42). Here was an application of his favored idea of unity in multeity, drawing equally on Neoplatonic and post-Kantian philosophies and differing markedly from Schelling and Steffens. His definition of life, similar to Schelling's, drew also on scholastic usage: "I define life as the *principle of individuation*, or the power which unites a given *all* into a *whole* that is presupposed by all its parts" (p. 42).[57] He instanced metals, for example, gold, as exponents of "the simplest form of unity, namely the unity of powers and properties" (p. 46). Crystals represented the next stage in the ascent, for in them was "a union, not of powers only, but of parts" (p. 47). Then came geological strata, residues and products of organic life that manifested "the tendencies of the Life of Nature to vegetation and animalization" (p. 48). This was precisely the argument presented in Steffens's *Beyträge*. Finally came "the present order of vegetable and animal Life" (p. 48), from the lowest vegetables to man.

Coleridge proceeded to give an account of powers, polarity, and the compass of nature, correlating the tetrad and pentad of the compass with the doctrine of three dimensions and with magnetism, electricity, and chemical process (pp. 50–70). All this was the prelude to a progressive zoology, in which the power of life ascended, as by a ladder (p. 70). The spaces between the rungs of the ladder symbolized the separation of different forms of life: animals from vegetables, man from apes, birds from fishes and insects. They were reconciled with "the equally evident continuity of the life of Nature" by an application of the principle of universal polarity. Polarization and differentiation were succeeded by a falling back and contraction, followed by a renewed act of polarization at a higher level. It is as if life had its own rhythm of systole and diastole, but with this difference, that the expansion of diastole, far from being a relaxation like that of the human heart, was a surge of ascent and creative power.

Successive geological formulations were products and residues of successive surges of life, schistose formations deriving from vegetation, whereas calcareous rocks were "the *caput mortuum* of animalized existence" (p. 71). A higher exertion of the power of life produced organized beings. Vegetation and animality were exponents of magnetism and electricity, manifested in their highest powers as reproduction, common to all living organisms, and irritability, common to all animals but lacking in most vegetables. These powers constituted the opposite poles of organic life, which existed in counterpoint, the higher vegetables being closest to the lower animals. Corals, for example, were plantlike and were also a manifest connection between the animal and mineral worlds. Coleridge's argument here, from the Wernerian formations to the culmination of terrestrial life in man, is based closely on Steffens's *Beyträge*, pp. 275–317, with occasional enrichment from his own zoological lore.

The ladder of life ascended from corals and polyps to molluscs, whose higher species possessed "the rudiment of nerves, as the first . . . impress and exponent of sensibility" (p. 73). Reproduction was, however, still predominant. Then, in preparation for the next class, came worms, in which the strife between irritability and reproduction was unresolved. Worms were succeeded by insects, whose variety of form and whose instincts showed that in them nature had rendered irritability predominant (p. 74). "THE INSECT WORLD IS THE EXPONENT OF IRRITABILITY" (p. 75). The growth of an exoskeleton, the beginning of a separation between the nervous and muscular system, and the appearance of compound eyes all served Coleridge as evidence that the tendency to individuation had advanced with the progressive dominance of irritability.

The next higher step corresponded to the class of fishes, where the skeleton was internal, but the bones were closer to gristle than to rigid mammalian bones (p. 79). Birds represented the next step, the "synthesis of fish and insect" (p. 83). Irritability was still the predominant power, but sensibility was now awake. And as for the progress of individuation, it was clear to every ornithologist "how large a stride has been now made by Nature . . . From a multitude of instances we select the most impressive, the power of sound, with the first rudiments of modulation! That all languages designate the melody of birds as singing . . . , demonstrates that it has been felt as, what indeed it is, a tentative and prophetic prelude of something yet to come" (pp. 83–4).

It was in the next step of the ascent of power, occupied by the mammals, that sensibility at last became predominant. The nervous system was most fully developed, the bones internal and solid. And at the top of the ladder, the crown of creation, was man (pp. 84–5):

> Man possesses the most perfect osseous structure, the least and most insignificant covering. The whole force of organic power has attained an inward and centripetal direction. He has the whole world in counterpoint to him, but he contains an entire world within himself. Now, for the first time at the apex of the living pyramid, it is Man and Nature, but Man himself is a syllepsis,[58] a compendium of Nature – the Microcosm!. . . In Man the centripetal and individualizing tendency of all Nature is itself concentred and individualized – he is a revelation of Nature! Henceforward, he is referred to himself, delivered up to his own charge; and he who stands the most on himself, and stands the firmest, is the truest, because the most individual, Man. In social and political life this acme is inter-dependence; in intellectual life it is genius. Nor does the form of polarity, which has accompanied the law of individuation up its whole ascent, desert it here. As the height, so the depth. The intensities must be at once opposite and equal. As the liberty, so must be the reverence for law. As the independence, so must be the service and submission to the Supreme Will! As the ideal genius and the originality, in the same proportion must be the resignation to the real world, the sympathy and the inter-communion with Nature! In the conciliating mid-point, or equator, does the Man live, and only by its equal presence in both its poles can that life be manifested![59]

Man, created in god's image, endowed with a living soul, self-consciousness, and self-government, was a microcosm of nature, and at the same time enshrined a world of spirit.[60] This meeting of two worlds, of nature and mind, rendered the one intelligible and the other accessible to reason in its near reaches and to faith in its more distant ones. When Coleridge wrote that we could know only what we were, he was not prescribing limits to knowledge; he was claiming the widest territories for reason and imagination.[61]

Conclusion

When Coleridge came to write his essays *On the Constitution of the Church and State*, he distinguished the Christian church from Christianity: "To the ascertainment and enucleation of the latter, of the great redemptive process which began in the separation of light from Chaos (*Hades, or the Indistinction*), and has its end in the union of life with God, the whole summer and autumn, and now commenced winter of my life have been dedicated. HIC Labor, HOC opus est, on which alone the author rests his hope, that he shall be found to have lived not altogether in vain."[62] That, certainly, was the program for the *Opus Maximum*. But the process beginning in "the separation of light from Chaos" and ending in "the union of life with God" comprehended the scientific and philosophical natural history of the cosmos. The history of nature was divinely ordained and determined, its scientific and spiritual aspects developing in a perfect complementarity and correspondence.

Coleridge possessed an imaginative sympathy with the enterprise of science. Because he had a unified vision of the natural world, he had no need to confine his inquiries within the traditional compartments respected by scientists. He was able to utilize the work of chemists, geologists, and physiologists, without insisting on the complete separation of their disciplines. Freedom from the limitations imposed by disciplinary boundaries enabled him to ask questions and suggest research in areas avoided by many of his contemporaries, and skepticism about conclusions based on atomism made him a trenchant critic of the cruder explanations of contemporary mechanistic science. Coleridge's philosophy and his knowledge of science repeatedly led him to point out inconsistencies, and to identify as significant problems avoided or ignored by contemporary natural philosophers.[63]

The distance between his view of nature and theirs was frequently immense. Even Davy wrote that nature "has no archetype in the human imagination."[64] We may be tempted to conceive a clear opposition between the empirical scientist and Coleridge, the romantic spectator of science philosophizing in Platonic fashion. The two seem unlikely to blend, and it is tempting to emphasize the dichotomy. The range of sources consulted by Coleridge should, however, caution us against too simplistic a distinction. Schelling is said to have castigated Ritter as an empiricist philistine; Ritter was in later years too careless of facts for Oersted's taste; Oersted knew that the English thought him too much of a German metaphysician, and Coleridge was indeed offended by his mingling the speculative with the empirical. Yet Ritter and Oersted provided at least empirical fodder for Davy, who still remained critical of German philosophy of nature. And Berzelius recoiled before Davy's hypotheses. The line dividing experiment from speculation was important for every scientist, but there was clearly no unanimity in England or Germany about where the line should be drawn or how it should be interpreted. Scientists did not fit into a clean dichotomy of respectable scientists on one hand and romantic ones on the other. Ritter and Oersted were empiricists influenced by *Naturphilosophie*. Coleridge's criticisms were sometimes more perceptive and less dogmatic than those of subsequent observers of science.[65]

His range in science was remarkable, especially in England. For a poet, it was unique. It draws our attention to the great variety of styles of science practised in the early nineteenth century, suggesting an unexpected openness in English science to continental developments, and a corresponding weakening of the assumed hegemony of the Royal Society. Natural philosophy in the early 1800s was ambitious and far from monolithic.

The comparison of Coleridge's approach to science with that of Hegel is informative. Hegel, almost exactly Coleridge's contemporary, produced an encyclopedia structurally similar to Coleridge's projected *Opus Maximum*, and comprising three sections, the *Logic*, the *Philosophy of Nature*, and the *Philosophy of Spirit*. Hegel, like Coleridge, was knowledgeable about science, and organized his material hierarchically, incorporating it into a comprehensive world view. The similarities between Hegel's and Coleridge's enterprise are prominent. It is therefore striking to find that Coleridge, embarking on Green's copy of Hegel's *Wissenschaft der Logik* (3 pts. in 2 vols., Nürnberg, 1812–16), read and annotated only the first part of volume one. His notes evince marked impatience: "The first 40 or 50 pages of the First Book seem to me bewilderment throughout from confusion of Terms."[66] Coleridge did not give Hegel the same sympathetic attention that he gave other, less gifted, writers. Why? One reason may be that Hegel had not sufficiently distanced himself from Schelling's implicit pantheism. Hegel was, besides, concerned more with the arrangement and organization of existing knowledge than with its development. He had a system. Coleridge, seeking a system, had a method. The distinction is fundamental. The more complete a system is, the more it describes and classifies knowledge, and the less it encourages new kinds of inquiry. Coleridge's life's work was one long inquiry. His refusal to limit himself initially to clear and distinct ideas often led him astray, but it kept his thought from ossification. He sought to refine concepts that gained definition through use and experience. His inquiry, like that of science itself, was living, generative, and far from abstract. His method, transcending the abstraction of his philosophy, was a major constituent of his intellectual vitality and of his continuing and major importance in our own imaginative life of the mind.

NOTES

Abbreviations used in the Notes

These abbreviations of Coleridge's works are used only for the editions cited here.

AR — Aids to Reflections. Edited by T. Fenby. Edinburgh, 1905.
BL — *Biographia Literaria*. Edited by John Shawcross. 2 vols. Oxford, 1907.
BM — British Museum.
CC — *The Collected Works of Samuel Taylor Coleridge*. General editor K. Coburn. Bollingen Series 75. London and Princeton, N.J., 1969–.
CL — *Collected Letters of Samuel Taylor Coleridge*. Edited by E. L. Griggs. 6 vols. Oxford, 1956–71. Reprint (vols. 1 & 2). 1966.
CM — Coleridge's marginalia. Coleridge frequently made notes not only on the margins of the books that he read, but also on the blank pages at the front and back of a volume. All such notes are referred to here as marginalia. In numbering the blank pages, I have followed the procedure that George Whalley is adopting in his edition of the marginalia. He calls all blank pages at the front and back of a volume, outside the printed portion of the volume, "fly pages." He numbers them outward from the first and last page bearing any print; front fly pages are indicated by a minus sign before the page number, rear fly pages by a plus sign. Thus, for example, the last footnote in this book refers to marginalia on pages −4−−1 of a work by Hegel, i.e., to front fly pages 4–1.
CN — *The Notebooks of Samuel Taylor Coleridge*. Edited by K. Coburn. Bollingen Series 50. 5 vols. New York, Princeton, N.J., and London, 1957–.
N — Coleridge autograph notebooks. Nos. 3½, 16–20, 21½, 22–30, 33–55, British Museum. No. 29, Berg Collection, New York Public Library. Nos. 56, 59–63, 65, K, L, M, N, P, Victoria College Library, University of Toronto. No. F°, Huntington Library.
PW — *Coleridge: Poetical Works*. Edited by E. H. Coleridge. Reprint. 2 vols. in 1. Oxford, 1973.
Scrofula — Victoria College Library S TR F13.1.
STC — Samuel Taylor Coleridge.

TL	*Hints towards the Formation of a More Comprehensive Theory of Life.* Edited by S. B. Watson. London, 1848.
VCL	Victoria College Library, University of Toronto.

Introduction

1 *The Table Talk and Omniana of Samuel Taylor Coleridge* (Oxford, 1917), Sept. 12, 1831.

2 *Coleridge and the Pantheist Tradition* (Oxford, 1969), p. xxxix. It will be clear from this Introduction that McFarland has set the scene for the essay that follows. It will also be clear that the premises I take from him, and find in STC, lead me to conclusions about the significance of STC's scientific thought entirely at odds with McFarland's dismissal of "Coleridge's scientific pretensions" as "now of no importance except as an expression of Coleridge's concern with the 'it is'" (ibid., p. 249).

3 Ibid., p. 222.

4 "Essay on Faith," in *The Complete Works of Samuel Taylor Coleridge*, ed. W. G. T. Shedd, (New York, 1884), *5*, 565.

5 *Collected Letters of Samuel Taylor Coleridge* (henceforth *CL*), ed. E. L. Griggs, 6 vols. (Oxford, 1956–71; reprint ed., vols. 1 and 2, 1966), *3*, 155.

6 *Aids to Reflection*, Conclusion, in *Works*, *1*, 363.

7 *The Philosophical Lectures of Samuel Taylor Coleridge*, ed. K. Coburn (London, 1949), p. 223; STC notebook (henceforth N) F°, f19^{v}.

8 *Philosophical Lectures*, p. 176.

9 Royal Institution of Great Britain, MS ca. 1808.

10 *Naturphilosophie* had its major impact on the life sciences, e.g. physiology and comparative anatomy, and on the sciences based on powers, e.g., galvanism. It was also significant in geology and chemistry. Alexander von Humboldt was perhaps the most distinguished critic of *Naturphilosophie* in science, and yet his great work, *Cosmos*, bears its clear imprint. It is remarkable that STC made such little use of Humboldt's writings; examples of that use are in British Museum Egerton MS 2800 ff190^{r}, 190^{v}, and BM Add MSS 47,519 f57^{v}.

11 *Studies in Romanticism*, *16*, no. 3 (Summer 1977), was devoted to science and romanticism. D. M. Knight, "The physical sciences and the romantic movement," *History of Science*, *9* (1970), 54–75; and W. D. Wetzels, "Aspects of natural science in German romanticism," *Studies in Romanticism*, *10* (1971), 44–59, were earlier surveys. M. C. Jacob, *The Newtonians and the English Revolution, 1689–1720* (Hassocks, Sussex, 1976), exemplifies the newer social history of science. Recent interest in phrenology typifies the relaxation of earlier positivism; see, for example, R. M. Young, *Mind, Brain, and Adaptation in the Nineteenth Century* (Oxford, 1970); and R. J. Cooter, "Phrenology: the provocation of progress," *History of Science*, *14* (1976), 211–34.

12 Vol. *3* (1973) and especially vol. *4* (in press) of *The Notebooks of Samuel Taylor Coleridge* (henceforth *CN*), ed. K. Coburn, Bollingen Series 50, 5 vols. (New York, Princeton, N.J., and London, 1957–), contain the bulk of STC's scientific notes. *Shorter Works and Fragments*, ed. H. J. Jackson, and J. R. de J. Jackson, to be published in *The Collected Works of Samuel*

Taylor Coleridge (henceforth *CC*), general ed. K. Coburn, Bollingen Series 75 (London and Princeton, N.J.), contains the *Theory of Life* and numerous scientific fragments.

13 STC, *Biographia Literaria*, ed. W. J. Bate and J. Engell, in *CC* (in press), chap. 18.

14 O. Barfield, *What Coleridge Thought* (Oxford, 1972), p. 131.

15 N28 f21ᵛ.

16 N28 ff38ᵛ–9.

17 *Pantheist Tradition*, pp. 111, 187.

18 "Coleridge and the romantic vision of the world," in *Coleridge's Variety: Bicentenary Studies*, ed. J. Beer (London, 1974), pp. 101–33, at 116 ff.

19 *What Coleridge Thought*, p. 36.

20 McFarland, *Pantheist Tradition*, chap. 4.

21 It is not appropriate to discuss here exaggerated charges of plagiarism. These have been most adequately dealt with in general in ibid., chap. 1, and in particular in the editors' introduction to the forthcoming *Biographia Literaria* (*CC*), the work in connection with which the charges have chiefly been made.

22 *Biographia Literaria* (*CC*), editors' introduction.

23 Ibid., chap. 13.

24 "Notes on Henry More", in *Works*, *5*, 113.

Chapter 1
Early years

1 STC's political involvements are described in John Colmer, *Coleridge, Critic of Society* (Oxford, 1959; reprint ed., 1967). Biographical information is given in James Dykes Campbell, *Samuel Taylor Coleridge: A Narrative of the Events of His Life* (London, 1894; reprint ed., Highgate, 1970); and L. Hanson, *The Life of S. T. Coleridge* (1938; reprint ed., New York, 1962).

2 Basil Willey, "Joseph Priestley and the Socinian moonlight," in his *The Eighteenth-Century Background* (Harmondsworth, 1962), pp. 162–95; P. M. Zall, "The cool world of Samuel Taylor Coleridge: Joseph Priestley, firebrand philosopher," *Wordsworth Circle*, *9* (1978), 64–70.

3 G. Watson, "The revolutionary youth of Wordsworth and Coleridge," *Critical Quarterly*, *18* (1976), 49–66; J. Beer, "The 'revolutionary youth' of Wordsworth and Coleridge: another view," *Critical Quarterly*, *19* (1977), 79–86.

4 Abridged ed., 3 vols. (Birmingham, 1790), *1*, xxiii.

5 *Coleridge: Poetical Works* (henceforth *PW*), ed. E. H. Coleridge, reprint ed., 2 vols. in 1 (Oxford, 1973), pp. 117, 123.

6 2d ed., 2 vols. (London, 1782). See *CN*, *1*, entry 64 n.

7 *PW*, *1*, 123. For STC and Hartley, see Richard Haven, "Coleridge, Hartley, and the mystics," *Journal of the History of Ideas*, *20* (1959), 477–94, and *Patterns of Consciousness: An Essay on Coleridge* (Amherst, Mass., 1969), passim, esp. 102–9, 142 ff. See also *Lectures 1795: On Politics and Religion*, ed. L. Patton and P. Mann, in *CC* (1971), index, for Priestley, Hartley, and Newton.

8 David Hartley, *Observations on Man, His Frame, His Duty, and His Expectations*, 2 vols. (London, 1791), *1*, pt. 1, pp. 5–8, 56, 73–83.
9 STC's marginalia (hereafter CM) to ibid., rear fly, about textus *1*, 81, copy in BM, cat. no. C.126.i.2.
10 Hartley, *Observations on Man*, *2*, 53, 56; *CL*, *1*, 137 (postmark Dec. 11, 1794).
11 G. Whalley, "Coleridge and Southey in Bristol, 1795," *Review of English Studies*, n.s. *1* (1950), 324–40, at 333.
12 T. H. Levere, "Dr. Thomas Beddoes and the establishment of his Pneumatic Institution," *Notes and Records of the Royal Society of London*, *32* (1977), 41–9.
13 Colmer, *Coleridge, Critic of Society*, p. 47 n. For Bristol as a radical intellectual center, see C. A. Weber, *Bristols Bedeutung für die englische Romantik* (Halle, 1935). J. D. Harris, "The Susquehannah trail: Coleridge's studies in the useful arts, natural history, and medicine," *Dissertation Abstracts International*, *36* (1976), 6704A, argues that the same principles motivated these studies and pantisocracy.
14 Beddoes edited a translation of *The Chemical Essays of Charles-William Scheele* (London, 1786), and also J. Mayow's *Chemical Experiments and Opinions Extracted from a Work Published in the Last Century* (Oxford, 1790).
15 *A Letter to Erasmus Darwin, M.D., on a New Method of Treating Pulmonary Consumption, and Some Other Diseases Hitherto Found Incurable* (Bristol, 1793), p. 40; Levere, "Dr. Thomas Beddoes."
16 Tom, son of Josiah Wedgwood, and Gregory, son of James Watt, were both consumptive. R. B. Litchfield, *Tom Wedgwood, the First Photographer* (London, 1903).
17 Postscript by R. L. Edgeworth to letter from Maria Edgeworth to Mrs. R. Clifton, July 21, 1793. Edgeworth, MSS on loan to Bodleian Library, Oxford; published in Marilyn Butler, *Maria Edgeworth: A Literary Biography* (Oxford, 1972), p. 110. For an account of the Lunar Society see Robert E. Schofield, *The Lunar Society of Birmingham* (Oxford, 1963).
18 Public Record Office, London, MSS H.O. 422.21, 422.22; Banks to G. Devonshire, Nov. 30, 1794, BM (Natural History) Dawson-Turner Collection f125; James Watt, Jr., to Ferriar, Dec. 19, 1794, Gibson-Watt MS 20, private collection of Major D. Gibson-Watt.
19 *The Watchman*, ed. Lewis Patton, in *CC* (1970), pp. 100, 6, 34.
20 Ibid.
21 "Extracts from Doctor Beddoes's correspondence with Dr. Darwin," in J. Stock, *Memoirs of the Life of Thomas Beddoes, M.D.*. . . (London and Bristol, 1811), app. 6, p. xxxv; Beddoes, *Letter to Erasmus Darwin*.
22 Darwin to Beddoes, Feb. 6, 1794, Bodleian Library MS Dep C 134; Darwin, *Zoonomia; or, The laws of Organic Life*, 2 vols. (London, 1794–6), and *The Botanic Garden, a Poem in Two Parts: Part 1, Containing the Economy of Vegetation; Part 2, The Loves of the Plants, with Philosophical Notes* (pt. 1, London, 1791; pt. 2, Lichfield, 1789); *CN*, *1*, 9, 10, 174; D. King-Hele, *Doctor of Revolution: The Life and Genius of Erasmus Darwin* (London, 1977); John Livingston Lowes, *The Road to Xanadu* (Boston, 1927), p. 473.
23 *CL*, *1*, 177.
24 *CL*, *1*, 305. Darwin's chief scientific poem, *The Botanic Garden*, referred

to works of Lavoisier, Priestley, Richard Kirwan, and Linnaeus, among others.

25 G. Watt to Davy, Jan. 28, 1799, in *Fragmentary Remains, Literary and Scientific, of Sir Humphry Davy, Bart., . . .*, ed. John Davy (London, 1858), p. 27.

26 Stock, *Life of Thomas Beddoes*, p. 301.

27 T. Beddoes, *A Memorial concerning the State of the Bodleian Library, and the Conduct of the Principal Librarian: Addressed to the Curators of That Library, by the Chemical Reader* (Oxford, 1787).

28 B. C. Nangle, *The Monthly Review, Second Series, 1790–1815: Indexes of Contributors and Articles* (Oxford, 1955).

29 *CL, 1*, 203; *CN, 1*, 9, 10, 13, 14, 32, 64, 93, 235, text and nn. Priestley's "*Opticks*" is *The History and Present State of Discoveries Relating to Vision, Light, and Colours* (London, 1772). Newton's works were edited by S. Horsley, 1779–85. The work by Beddoes and J. Watt is *Considerations on the Medicinal Use and Production of Factitious Airs* (Bristol, 1795). J. Haygarth's work is "Description of a Glory," *Memoirs of the Literary and Philosophical Society of Manchester, 3* (1790), 463–7. Lowes, *The Road to Xanadu*, pp. 14, 457, cites the entry subsequently published as *CN, 1*, 32, a reference to John Hunter's article in the *Philosophical Transactions of the Royal Society of London*, 77 (1787), 371–450. See also G. Whalley, "The Bristol Library borrowings of Southey and Coleridge, 1793–8," *Library, 4* (1949), 114–31.

30 *CL, 1*, 260. M. Schofield, "Southey, Coleridge and company," *Contemporary Review, 230* (1977), 148–50, looks at the poets as dilettantes in chemistry. So they were in 1796.

31 *PW, 1*, 132, 113–14.

32 *CL, 1*, 279, 349.

33 *The Literary Remains of Samuel Taylor Coleridge*, ed. H. N. Coleridge, 4 vols. (London, 1836–9), *1*, 263.

34 Beddoes, *State of the Bodleian Library*, pp. 11, 15, 18.

35 *CL, 1*, 209; Mrs. Henry Sandford, *Tom Poole and His Friends*, 2 vols. (London, 1888).

36 *CL, 1*, 372, 256. For an account of STC's thoughts on education, see W. Walsh, *Coleridge: The Work and the Relevance* (London, 1967).

37 Quoted in Campbell, *Coleridge*, p. 64.

38 *CL, 1*, 320. Joseph Cottle's own account of his relations with STC is his *Reminiscences of Samuel Taylor Coleridge and Robert Southey* (London, 1847; reprint ed., Highgate, 1970).

39 *CL, 1*, 413; Campbell, *Coleridge*, 91–2. STC pays tribute to Tom Wedgwood in *The Friend* ed. B. Rooke, in *CC*, 2 vols. (1969), *1*, 147.

40 *CL, 1*, 444, 284.

41 *CL, 1*, 475.

42 T. Beddoes, review of J. F. Blumenbach, *De generis humani varietate nativa* 3d. ed., in *Monthly Review*, 2d ser. *21* (1796), 9. Beddoes's edition of Mayow, *Chemical Experiments*, p. xxxviii, praises "Blumenbach Instit. Physiologicae . . . Goettingae, 1787."

43 *CL, 1*, 494; STC, *Biographia Literaria*, ed. George Watson (London and New York, 1965), p. 116; *Friend, 1*, 494 n; *CN, 1*, 1657, *3*, 3744; STC,

Hints towards the Formation of a More Comprehensive Theory of Life (hereafter *TL*), ed. S. B. Watson (London, 1848), p. 83. See Chap. 7 of this book.

44 *CN, 1,* 388, re *Allgemeine Literatur Zeitung,* 1799, pt. 1, pp. 377–82, 385–97, 401–6, 409–14, 417–20, 426–9, 433–7, 441–60, 466–70 (see Chap. 7 of this book); *CL, 1,* 494. M. Crosland, *The Society of Arcueil* (London, 1967), pp. 32–6, discusses work on beet sugar.

45 C. Carlyon, *Early Years and Late Reflections,* 4 vols. (London, 1836–58), *1,* 32; Edith J. Morley, "Coleridge in Germany (1799)," in *Wordsworth and Coleridge: Studies in Honor of George McLean Harper,* ed. E. L. Griggs (Princeton, N.J., 1939), pp. 220–36; Cambridge University Library, Greenough MSS Add. 7918 (14); *CN, 1,* 1657, *2,* 1864, 2014; *CL, 1,* 521.

46 *CN, 1,* 430, 431, and nn. Stephen Prickett, *Coleridge and Wordsworth: The Poetry of Growth* (Cambridge, 1970), discusses the significance of the glory for Coleridge.

47 I. A. Richards, "Coleridge: his life and work," in *Coleridge: A Collection of Critical Essays,* ed. K. Coburn (Englewood Cliffs, N.J., 1967), pp. 12–31, at 19; K. Coburn, ed., *Inquiring Spirit: A New Presentation of Coleridge* (London, 1951; new ed., Toronto, 1979), p. 234; *CL, 1,* 502–3.

48 *Friend, 1,* 471. Cf. J. Beer, *Coleridge's Poetic Intelligence* (London and Basingstoke, 1977); and J. E. Stewart, "Samuel Taylor Coleridge," *Dissertation Abstracts International, 38* (1977), 291A.

49 *CL, 1,* 518; *CN,1,* 433. His spelling in the notebooks attempts to be phonetic.

50 *CL, 1,* 519; Carlyon, *Early Years,* p. 187.

51 Carlyon, *Early Years,* pp. 185, 187; *CL, 1,* 519. For STC's use of Blumenbach, see Chap. 7 of this book.

52 *The Collected Works of Sir Humphry Davy, Bart. . . .,* ed. John Davy, 9 vols. (London, 1839), *1,* 13, reprinted in Harold Hartley, *Humphry Davy* (London, 1966), p. 12; *CL, 1,* 320.

53 Davy wrote several poems for Southey: *The Annual Anthology,* 2 vols. (Bristol, 1799–1800), *1,* 93–9, 120–5, 172–6, 179–80, 281–6, *2,* 293–6.

54 J. Z. Fullmer, "The poetry of Humphry Davy," *Chymia, 6* (1960), 102–26.

55 Davy, *Fragmentary Remains,* pp. 34 ff.; R. Sharrock, "The chemist and the poet: Sir Humphry Davy and the preface to the *Lyrical Ballads*" *Notes and Records of the Royal Society of London, 17,* (1962), 57–76; *CL, 1,* 549.

56 *Works, 1,* 66–7. H. W. Piper, *The Active Universe* (London, 1962), explores the attitudes implicit in this entry.

57 T. Beddoes, *Notice of Some Observations Made at the Medical Pneumatic Institution* (Bristol, 1799), p. 40; S. R. Hoover, "Coleridge, Humphry Davy, and Some Early Experiments with a Consciousness-Altering Drug," *Bulletin of Research in the Humanities, 81* (1978), 9–27.

58 J. A. Paris, *The Life of Sir Humphry Davy . . . ,* 2 vols. (London, 1831), *1,* 74–5; K. Coburn, "Coleridge, a bridge between science and poetry: reflections on the bicentenary of his birth," *Proceedings of the Royal Institution of Great Britain, 46* (1973), 45–63, at 48.

59 CM, *John Barclay His Argenis,* trans. R. Le Grys and T. May (London, 1629), to be published in G. Whalley's edition of STC's *Marginalia* for *CC.*

60 *CL, 1,* 549, 559. Wordsworth was the superior being.

61 *CL*, *1*, 557.
62 T. Beddoes, ed., *Contributions* . . . (Bristol, 1799), pp. 5–147.
63 A. L. Lavoisier, *Traité élémentaire de chimie*, 2 vols. (Paris, 1789), trans. as *Elements of Chemistry* by Robert Kerr (Edinburgh, 1790), p. 175. Locke and Condillac are prominent in Lavoisier's preface. For the English reception of Lavoisier's chemistry, see J. R. Partington, *A History of Chemistry*, vol. 3 (London, 1962).
64 *CL*, *1*, 538.
65 Beddoes, *Contributions*, pp. 221–3.
66 Davy, *Works*, *2*, 73–4.
67 H. Davy, "An essay on heat, light, and the combinations of light," in Beddoes, *Contributions*, pp. 4–147, at 144–7.
68 *CL*, *1*, 588, 590; J. F. Blumenbach, *Handbuch der Naturgeschichte* (Göttingen, 1799). See entry in Index 2 of *CN*, *1*.
69 *CL*, *1*, 605 (July 15 [16], 1800).
70 *CL*, *1*, 494; *CN*, *1*, 662.
71 *CN*, *1*, 879 n. The original notebook entry is in Greek; the translation is given in the editor's note. Religion was the chief fact in STC's intellectual life – including his assessment of science. For a general statement of this thesis, see Basil Willey, *Samuel Taylor Coleridge* (London, 1972).
72 CM, *The Works of Jacob Behmen [Boehme], the Teutonic Theosopher . . . to Which Is Prefixed, the Life of the Author* . . . , ed. G. Ward and T. Langcake, 10 pts. in 4 vols. (London, 1764–81), *1*, pt. 1, pp. 39–40, copy in BM, cat. no. C.126.k.1.
73 Quoted in Anne Treneer, *The Mercurial Chemist: A Life of Sir Humphry Davy* (London, 1963), p. 61.
74 *CL*, *1*, 557; *Friend*, *1*, 471; STC, *Literary Remains*, *1*, 223; *CL*, *5*, 309, *1*, 630–1; Davy, *Fragmentary Remains*, p. 81 n; J. Davy, *Memoirs of the Life of Sir Humphry Davy, Bart., LL.D., F.R.S.*, 2 vols. (London, 1836), *1*, 390.
75 *CL*, *1*, 605–6; H. Davy, *Researches Chemical and Philosophical; Chiefly Concerning Nitrous Oxide, or Dephlogisticated Air, and Its Respiration* (London, 1800); *CL*, *1*, 611.
76 *CL*, *1*, 611, 617; Sharrock, "The chemist and the poet."
77 *CL*, *1*, 623, 630. Davy's publications on electrochemistry are listed in June Z. Fullmer, *Sir Humphry Davy's Published Works* (Cambridge, Mass., 1969). T. H. Levere, *Affinity and Matter: Elements of Chemical Philosophy, 1800–1865* (Oxford, 1971), chap. 2, describes these researches. Introductions to dynamical chemistry are provided by David M. Knight, "Steps towards a dynamical chemistry," *Ambix*, *14* (1967), 179–97, *Atoms and Elements* (London, 1967), and *The Transcendental Part of Chemistry*, (Folkestone, 1978). See Chap. 6 of this book.
78 Davy to STC, Nov. 26, 1800, Pierpont Morgan Library, MS MA 1857 no. 11; *CL*, *1*, 648, *2*, 670–2, 734.
79 *CL*, *2*, 708 (Locke was the perfect "Little-ist"); *CN*, *1*, 857; Paul Deschamps, *La Formation de la pensée de Coleridge, 1772–1804*, Études Anglaises no. 15, (Grenoble, 1964), pt. 3.
80 G. A. Rogers, "Locke's *Essay* and Newton's *Principia*, *Journal of the History of Ideas*, *39* (1978), 217–32, reexamines this simplistic assumption.
81 *CL*, *2*, 709.
82 *CL*, *2*, 712; *PW*, pp. 242, 366.

83 *CL*, *2*, 727; William Herschel, "Experiments on the refrangibility of the invisible rays of the sun," *Philosophical Transactions*, *90* (1800), 284–92; Davy, "Heat, light, and the combinations of light"; and "An essay on the generation of phosoxygen, or oxygen gas: and on the causes of colours of organic beings," in Beddoes, *Contributions*, pp. 151–98, 199–205.
84 Treneer, *The Mercurial Chemist*, p. 37.
85 *CN*, *1*, 1034.
86 Davy to John King, June 22, 1801, Bristol City Archives MS 32688/31; *CL*, *2*, 745–6, 751, 767–8.
87 *CL*, *2*, 776–9, 782; Davy, *Works*, *2*, 307–26, 327–436. Cornell University Library, Wordsworth Collection, has the copy that Davy sent to STC of *A Syllabus of a Course of Lectures on Chemistry, Delivered at the Royal Institution of Great Britain* (London, 1802).
88 "Mr. Davy's lectures on chemistry," *Journal of the Royal Institution of Great Britain*, *1* (1802), 109–12. See Fullmer, *Davy's Published Works*, p. 35.
89 Davy, *Works*, *2*, 323; Paris, *Life*, *1*, 92. STC's notes on Davy's lectures are published in *CN*, *1*, 1098, 1099.
90 STC, *Aids to Reflection* (hereafter *AR*), ed. T. Fenby (Edinburgh, 1905), pp. 181–2. There is a valuable discussion of STC's use of metaphor in M. H. Abrams, *The Mirror and the Lamp: Romantic Theory and the Critical Tradition* (Oxford, 1953). See also Walsh, *Coleridge*, p. 164.
91 BM Add MS 47,521 ff31–31v.
92 BM Add MS 47,524 f131v; STC, *Opus Maximum*, Victoria College Library (henceforth VCL), University of Toronto, S MS 29, *1*, ff38, 32.
93 *CN*, *3*, 3312. A. Thackray, *Atoms and Powers* (Cambridge, Mass., 1970), describes the eighteenth-century Newtonian basis for STC's model of chemistry working through affinities as powers. For dynamical chemistry, see Chap. 6 of this book and n. 77 to this chapter.
94 *Opus Maximum*, VCL S MS 29, *1*, passim; notebook entitled "Marginalia Intentionalia," Berg Collection, New York Public Library, fl9; VCL S MS 5.
95 VCL S MS 29, *1*, f38; Royal Institution, Davy papers, MS 20c.
96 Dorothy Emmet, "Coleridge on powers in mind and nature," in *Coleridge's Variety: Bicentenary Studies*, ed. John Beer (London, 1974), pp. 166–82.
97 *CN*, *1*, 1099. STC's sense and use of color is similarly and strikingly evident in his "Ancient Mariner."
98 *CL*, *2*, 927, 1028.
99 *CL*, *2*, 1942, 1046–7, 1071, 1080; *CN*, *2*, 1954, 1872, 1859. The quotation is from *CN*, *2*, 1855.
100 *CN*, *2*, 1864. A. Hayter, *Opium and the Romantic Imagination* (London, 1968), chap. 9, discusses STC and opium, and Hayter has a useful bibliography on this topic on p. 360. See also M. Lefebure, *Samuel Taylor Coleridge: A Bondage of Opium* (1974); CL, 2, 1059; and A. Hayter, *A Voyage in Vain: Coleridge's Journey to Malta in 1804* (London, 1973).
101 Dove Cottage Trust, Grasmere, MS 72.
102 Royal Institution, Davy MS 13c.
103 For background see Hartley, *Humphry Davy*; Levere, *Affinity and Matter*, chap. 2; and C. A. Russell, "The electrochemical theory of Humphry Davy," *Annals of Science*, *15* (1959), l–13, 15–25, *19* (1963), 255–71.

104 A Volta, "On the electricity excited by the mere contact of conducting surfaces of different kinds," *Philosophical Transactions, 90,* (1800), 403–31; J. Priestley to H. Davy, Oct. 31, 1801, in Davy, *Fragmentary Remains,* pp. 51–3; T. Poole to J. Wedgwood, Jan. 15, 1805, Wedgwood papers, Keele University, MS 846–5; Levere, *Affinity and Matter,* p. 35.
105 Davy, *Works, 2,* 162, 309 ff.; Paris, *Life, 1,* 109–10.
106 *CL, 1,* 630; Davy, *Works, 2,* 201, 208.
107 Davy to STC, Nov. 25, 1800, Pierpont Morgan Library, MS MA 1857 no. 11; *CL, 2,* 664.
108 Davy, *Works, 2,* 189–209; *CL, 2,* 734, 726–7.
109 Levere, *Affinity and Matter,* p. 36; Davy, *Fragmentary Remains,* p. 115. Cf. STC to Davy, *CL, 2,* 1102.
110 "The Bakerian Lecture, on some chemical agencies of electricity," *Philosophical Transactions, 97* (1807), 1–56.
111 "The Bakerian Lecture, on some new phenomena of chemical changes produced by electricity," *Philosophical Transactions, 98* (1808), 1–44; Levere, *Affinity and Matter,* p. 41; Hartley, *Humphry Davy,* pp. 50–63.
112 Davy, *Fragmentary Remains,* p. 90, *variatim; CL, 3,* 41.
113 Beddoes to Watt, June 20, 1807, in James P. Muirhead, *The Origins and Progress of the Mechanical Inventions of James Watt,* 2 vols. (London, 1854), *2,* 306; Beddoes to Joseph Black, Feb. 23, 1788, Black MSS, Edinburgh University Library; J. J. Berzelius, *Traité de chimie,* trans. A. J. L. Jourdain and M. Esslinger, 8 vols. (Paris, 1829–33), *1,* 164; H. Brougham, "Davy's *Bakerian Lecture,*" *Edinburgh Review, 11* (1808), 390–8, at 398.
114 *CL, 3,* 38.
115 Ibid.
116 *CL, 3,* 135, 144–5, 170–3; CM, Behmen [Boehme], *Works, 1,* pt. 1, p. 40, copy in BM.
117 Levere, *Affinity and Matter,* chap. 2; Knight, *Atoms and Elements,* passim; R. Ziemacki, "Humphry Davy and the conflict of traditions in early 19th-century British chemistry," Ph.D. diss., Cambridge University, 1975.

Chapter 2
Surgeons, chemists, and animal chemists

1 B. Willey, *Samuel Taylor Coleridge* (London, 1972), chap. 10, pp. 114–15.
2 *CN, 3,* 3570.
3 *CN, 3,* 3476.
4 See Chap. 7.
5 *CN, 3,* 3575; *Theory of Life,* passim; *CN, 3,* 3811, 3632. STC's account of life is discussed more fully in chap. 7. J. F. Blumenbach, *Ueber den Bildungstrieb und das Zeugungsgeschäft* (Göttingen, 1781; 3d ed., 1791), was a turning point in the discussion of life as a power. German philosophers – e.g., F. W. J. von Schelling, *Von der Weltseele* (Hamburg, 1809), p. xliv – incorporated this into their ideas about productivity in nature (see Chap. 4 of this book, the section entitled "The history of nature and the productivity of nature"). STC, however, was drawing not only on recent German thought, but on Platonic and subsequent Scholastic thought, as in *The Friend,* no. 5, Sept. 14, 1809, in *Friend, 2,* 75.

6 *CN*, *3*, 3744 (Mar. 1810) and n.

7 *CN*, *3*, passim. Coleridge was rereading Kant all his life; a full list of his known readings of Kant will be published by G. Whalley in the forthcoming edition of the *Marginalia*, *CC*. For Coleridge's reading of Fichte, see *CN*, *3*, 3276 n. In this context, the most pertinent works of Schelling are *Ideen zu einer Philosophie der Natur* (Landshut, 1803), *System des transcendentalen Idealismus* (Tübingen, 1800), and *Einleitung zu seinem Entwurf eines Systems der Naturphilosophie* (Jena and Leipzig, 1799), all quoted in *Biographia Literaria*, notably in chap. 12 (see *Biographia Literaria* in *CC*, forthcoming). Also important is Schelling's "Allgemeine Deduction des dynamischen Processes," *Zeitschrift für spekulative Physik*, *1* (1800), pt. 1, pp. 100–36, pt. 2, pp. 1–87 (see Chap. 5 of this book). The later Schelling is represented by *Philosophie und Religion* (Tübingen, 1804). For Steffens, see n. 31 to this chapter. For J. N. Tetens, see *Philosophische Versuche über die menschliche Natur und ihre Entwicklung* (Leipzig, 1775–6), discussed in *CN*, *2*, 2375 n.

8 G. Marcel, *Coleridge et Schelling* (Paris, 1971).

9 Michaud's *Biographie Universelle*; *Allgemeine Deutsche Biographie*.

10 *CL*, *3*, 461, 422 (related interests appear in *CN*, *3*, 3606–7 [1809]), *4*, 664.

11 *CL*, *4*, 666.

12 *CL*, *4*, 730, 738. Cf. *CN*, *3*, 4307.

13 T. McFarland, *Coleridge and the Pantheist Tradition* (Oxford, 1969).

14 *CN*, *3*, 4012 (Nov. 1810).

15 CM at end of *Critik der reinen Vernunft* (Leipzig, 1799), BM cat. no. C.126.i.9, transcribed in VCL MS BT 22, p. 30. For example, Coleridge checked Steffens against W. T. Brande, *A Manual of Chemistry* (London, 1819).

16 Willey, *Coleridge*; A. Loades, "Coleridge as theologian: some comments on his reading of Kant," *Journal of Theological Studies*, *29* (1978), 410–26; D. McVeigh, "Coleridge and political imagination," *Dissertation Abstracts International*, *38* (1977), 3518A–19A.

17 STC, *Biographia Literaria* (henceforth *BL*), ed. John Shawcross, 2 vols. (Oxford, 1907), *1*, 1.

18 *CL*, *4*, 767.

19 *BL*, *1*, 11 (chap. 1), 38–9 (chap. 3), 74 (chap. 6), 85 (chap. 7), 59 (chap. 5).

20 *BL*, *1*, 85–6 (chap. 7).

21 *BL*, *1*, 44 (chap. 3).

22 *BL*, *1*, 87–8 (chap. 7).

23 *BL*, *1*, 98, 102–3 (chap. 9), 162–3 (chap. 12). Boehme had been known to him since schooldays.

24 *BL*, *1*, 170–6 (chap. 12). Coleridge here draws closely from the opening 3 pages of Schelling's *System des transcendentalen Idealismus*.

25 *BL*, *1*, 202 (chap. 13); L. C. Knights, "Ideas and symbol: some hints from Coleridge," in *Coleridge: A Collection of Critical Essays*, ed. K. Coburn (Englewood Cliffs, N.J., 1967), pp. 112–22. Cf. *Friend*, *1*, 471. For STC's sources and fellow inquirers in the philosophy of nature and imagination, see *Biographia Literaria*, ed. W. J. Bate and J. Engell, in *CC* (in press), intro. and notes to chaps. 12 and 13. See also T. J. Corrigan, "*Biographia*

Literaria and the language of science," *Journal of the History of Ideas, 41* (1980), 399–419.

26 STC, *Literary Remains, 2*, 54 (concerning Shakespeare lectures, 1818); *BL, 2*, 19–20 (chap. 15); T. H. Levere, "S. T. Coleridge: a poet's view of science," *Annals of Science, 35* (1978), 33–44.

27 VCL S MS F13.4; *CL, 2*, 706, 916, *4*, 575, 688, 690; *CN, 3*, 4328; H. Jackson, Introduction to *Theory of Life, CC*; H. Nidecker, "Praeliminarien zur Neuausgabe der Abhandlung über die Lebenstheorie (Theory of Life) von Samuel Taylor Coleridge," Ph.D. diss., University of Basel, 1924; G. N. G. Orsini, *Coleridge and German Idealism* (Carbondale, Ill., 1969), pp. 232–3; J. H. Haeger, "Coleridge's 'Bye Blow': the composition and date of *Theory of Life*," *Modern Philology, 74* (1976), 20–41. See also Chap. 7 of this book.

28 STC, *Essay on Scrofula* (henceforth *Scrofula*) VCL S TR F13.1, p. 18.

29 Details of the competition are given in *London Medical and Physical Journal, 36* (1816), 254.

30 *CL, 4*, 688–90, *5*, 49–50; H. J. Jackson, "Coleridge on the king's evil," *Studies in Romanticism, 16* (1977), 337–47.

31 A. D. Snyder, *Coleridge on Logic and Learning* (New Haven, Conn., 1929), pp. 16–25. For Schelling, see n. 7 to this chapter. The Steffens works were *Grundzüge der philosophischen Naturwissenschaft* (Berlin, 1806) and *Beyträge zur innern Naturgeschichte der Erde* (Freiburg, 1801): STC's copies of these two works were bound together in 1815 (see *CN, 3*, 4223, 4226, and nn). *CL, 5*, 49–50, sketches STC's awareness of the Lawrence–Abernethy debate. See also O. Temkin, "Basic science, medicine, and the romantic era," *Bulletin of the History of Medicine, 37* (1963), 97–129; and J. Goodfield-Toulmin, "Some aspects of English physiology," *Journal of the History of Biology, 2* (1969), 283–320.

32 Biographical information about Green is in J. H. Green, *Spiritual Philosophy: Founded on the Teaching of the Late Samuel Taylor Coleridge*, ed., with a memoir of the author's life, by John Simon, 2 vols. (London, 1865).

33 Royal College of Surgeons of England, MS 67.b.11, notes from Green's lectures taken by W. H. Clift and probably Richard Owen, 1824. See also RCS MS 42.a.19, T. E. Bryant's notes from Green's lectures of 1827.

34 J. H. Green, *Vital Dynamics* (London, 1840); see esp. pp. 33–42. Green also repeatedly quotes STC in his second Hunterian Oration, *Mental Dynamics* (London, 1847).

35 Green acknowledges Karl Gustav Carus, Christian Wolff, and Johann Friedrich Meckel. For STC and German physiology, see Chap. 7. See also Green, *Vital Dynamics*, pp. 37 n, xix. Green's statement of his ignorance of the *Theory of Life* until he saw it at the press is in VCL S MS F13.5. Heather Jackson told me about this evidence.

36 These revisions appear to be in Sara Coleridge's hand, where STC's had been.

37 *CN, 3*, 4328.

38 *CL, 4*, 369–72; Rees's *Cyclopaedia, 20* (1819). The article on life is identified as Lawrence's in P. C. Ritterbush, *Overtures to Biology: The Speculations of Eighteenth-Century Naturalists* (New Haven, Conn., 1964), p. 247.

39 See, e.g., N28 ff2v–3, and entries throughout N27, N28, N29.
40 "Monologues by the late Samuel Taylor Coleridge (no. I): on life," *Fraser's Magazine, 12* (1835), 493–6, at 494.
41 Jackson, "Coleridge on the king's evil."
42 *Scrofula*, p. 1. STC surely considered himself prone to the disease. T. Beddoes, *Hygeia; or, Essays Moral and Medical on the Causes Affecting the Personal State of Our Middling and Affluent Classes*, 3 vols. (Bristol, 1802–3), *2*, essay 6, "Essay on Scrophula," pp. 20–3, describes the physical and intellectual characteristics of those "of the scrophulous temperament," in a fashion that embraces STC in his self-portrait to John Thelwall, *CL, 1*, 259–60.
43 H. Jackson, Introduction to *Shorter Works and Fragments, CC*; *Scrofula*, passim.
44 *Scrofula*, pp. 14, 18.
45 Ibid., p. 15r. See *Essay on Scrofula*, in *Shorter Works and Fragments, CC* (forthcoming); and *TL*, pp. 59–65.
46 STC, *Philosophical Lectures*, pp. 28, 24.
47 E. Home, *Lectures on Comparative Anatomy*, 6 vols. (London, 1814–28), *1*, 6–23; John Hunter, *A Treatise on the Blood, Inflammation and Gunshot Wounds, . . . to Which Is Prefixed a Short Account of the Author's Life, by His Brother-in-law Everard Home* (London, 1794), pp. xxxviii–xlv.
48 *Friend*, 1, 474.
49 It should, however, be noted that Abernethy at the beginning of *An Enquiry into the Probability and Rationality of Mr. Hunter's Theory of Life . . .* (London, 1814), p. 79, refers to concepts implicit in the Hunterian collection rather than in Hunter's works.
50 Hunter, *Treatise on the Blood*, p. 77.
51 CM, Behmen [Boehme], *Works 1*, pt. 1, pp. 249–51, copy in BM; Hunter, *Treatise on the Blood*, p. 78.
52 See, e.g., G. Berkeley, *Alciphron* (London, 1732), *3*, §1. To unravel the eighteenth-century meanings of "principle" would be to make a seminal contribution to intellectual history.
53 J. E. McGuire, "Force, active principles, and Newton's invisible realm," *Ambix, 15* (1968), 154–208; P. M. Heimann and J. E. McGuire, "Newtonian forces and Lockean powers: concepts of matter in eighteenth-century thought," *Historical Studies in the Physical Sciences, 3* (1971), 233–306; A Thackray, *Atoms and Powers* (Cambridge, Mass., 1970); R. E. Schofield, *Mechanism and Materialism* (Princeton, N.J., 1970); Ernan McMullin, *Newton on Matter and Activity* (Notre Dame, Ind., 1978).
54 Abernethy, *Enquiry*, pp. 8, 17; J. Goodfield-Toulmin, "Aspects of English physiology," esp. pp. 293–5.
55 Abernethy, *Enquiry*, pp. 42, 49–50.
56 *Edinburgh Review* (1814), 384–98, 384, quoted in Goodfield-Toulmin, "Aspects of English physiology," p. 286; J. Adams, *An Illustration of Mr. Hunter's Doctrine Particularly Concerning the Life of the Blood, in Answer to the Edinburgh Review of Mr. Abernethy's Lectures* (London, 1814). His presentation copy to STC is in the BM.
57 W. Lawrence, *An Introduction to Comparative Anatomy and Physiology . . .* (London, 1816), v, 3, 92, 120, 150, 161, 170–4; X. Bichat, *Physiological*

Researches on Life and Death, trans. F. Gold (London, n.d.), p. 21, reviewed in *London Medical and Physical Journal*, *35* (1816), 41–55, 125–41; W. R. Albury, "Experiment and explanation in the physiology of Bichat and Magendie," *Studies in the History of Biology*, *1* (1977), 47–131.

58 J. Abernethy, *Physiological Lectures, Exhibiting a General View of Mr. Hunter's Physiology* (London, 1817), pp. 27, 38, 42.

59 W. Lawrence, *Lectures on Physiology, Zoology, and the Natural History of Man* . . . (London, 1819), pp. 5, 7, 8, 12–13.

60 Goodfield-Toulmin, "Aspects of English physiology," pp. 283–4.

61 *TL*, p. 17; *CL*, *4*, 688.

62 *TL*, pp. 21–2. This is also printed in *Philosophical Lectures*, pp. 355–6.

63 Lawrence, *Introduction to Comparative Anatomy*, p. 147 (ref. supplied by H. Jackson); "Abernethy on the Vital Principle," *Edinburgh Review*, *23* (1814), 384–98, at 386; *TL*, p. 24.

64 *TL*, p. 25.

65 See Chap. 4 of this book. See also I. Kant, *Logic*, trans. R. S. Hartman and W. Schwarz (Indianapolis, 1974), §106, p. 144; *TL*, p. 36; and Lawrence, *Introduction to Comparative Anatomy*, p. 167. Goodfield-Toulmin, "Aspects of English physiology," p. 284, points out that Lawrence was also indebted to Scottish philosophy, although she sees no conflict between the influence of Hume and that of Thomas Brown.

66 See n. 31 to this chapter. Cf. Steffens, *Beyträge*, pp. 37–8; and *TL*, pp. 31–4. STC's history of science will be discussed in Chap. 3.

67 *Philosophical Lectures*, p. 354 (see also Lawrence, *Introduction to Comparative Anatomy*, p. 115), p. 348; Hunter, *Treatise on the Blood*, p. 78.

68 *TL*, pp. 60–1; Lawrence, *Introduction to Comparative Anatomy*, pp. 168–71.

69 *TL*, pp. 63–6. Cf. *CN*, *2*, 2319 (Dec. 12, 1804).

70 *TL*, p. 65; "Abernethy, Lawrence, &c. *on the Theories of Life*," *Quarterly Review*, *22* (1819), 1–34.

71 E.g., N27 f55[v].

72 N27 f93[v].

73 N27 f93; *CL*, *4*, 809. Cf. N27 f92[v]; *Table Talk*, June 29, 1833.

74 *Friend*, *1*, 473–5.

75 Everett Mendelsohn, *Heat and Life: The Development of the Theory of Animal Heat* (Cambridge, Mass., 1964); J. J. Berzelius, "On the analysis of organic bodies," *Annals of Philosophy*, *4* (1814), 323–31, 400–9, *5* (1815), 93–101, 174–84, 260–75.

76 Home, *Lectures on Comparative Anatomy*, *5*, 14–15.

77 *CL*, *4*, 688–90.

78 *Friend*, *1*, 16.

79 *CL*, *4*, 809. The essays on method will be discussed in Chap. 4.

80 *Friend*, *1*, 94 and n.

81 Ibid., p. 97.

82 Ibid., p. 104.

83 *AR*, p. 363.

84 O. Barfield, *What Coleridge Thought* (Oxford, 1972), chaps. 8 and 9.

85 *Friend*, *1*, 156–8; Milton, *Paradise Lost*, bk. 5, lines 486–7.

86 *Friend*, *1*, 185.

87 Ibid., p. 449.
88 *AR*, pp. xvii–xviii.

Chapter 3
Two visions of the world

1 *CL*, *4*, 768.
2 *BL*, *1*, 135 (chap. 10).
3 *CN*, *1*, 174 (1796). Space between "Pride" and "Proud" is in the original.
4 BM Egerton MS 2801 ff212–15.
5 A. Thackray, *Atoms and Powers* (Cambridge, Mass., 1970); R. E. Schofield, *Mechanism and Materialism* (Princeton, N.J., 1970); T. H. Levere, *Affinity and Matter: Elements of Chemical Philosophy, 1800–1865* (Oxford, 1971); J. E. McGuire, "Atoms and the 'analogy of nature': Newton's third rule of philosophizing," *Studies in the History and Philosophy of Science, 1* (1970), 3–58; P. M. Heimann and J. E. McGuire, "Newtonian forces and Lockean powers: concepts of matter in eighteenth-century thought," *Historical Studies in the Physical Sciences, 3* (1971), 233–306; Ernan McMullin, *Newton on Matter and Activity* (Notre Dame, Ind., 1978).
6 I. Newton, *Philosophiae Naturalis Principia*, 3d ed. (London, 1726), pp. 387–9, reprinted in *Isaac Newton's . . . Principia*, ed. A. Koyré and I. B. Cohen (Cambridge, 1972), *2*, 550–5; Maurice Mandelbaum, *Philosophy, Science, and Sense Perception: Historical and Critical Studies* (Baltimore, 1964).
7 J. Priestley, *Disquisitions on Matter and Spirit*, 2d ed., 2 vols. (London, 1782), was the culmination of this tradition.
8 Priestley, *Matter and Spirit*, *1*, 43; *AR*, p. 361.
9 BM Egerton MS 2801 f63.
10 *CN*, *1*, 203 and n; BM Egerton MS 2801 f215; CM, *The Birds of Aristophanes*, trans. H. F. Cary, (London, 1824), pp. 64–5, copy in Harvard University Library. I have found no discussion in STC of the Hutchinsonians; he would have dismissed them as material theists. Their views are presented in C. B. Wilde, "Hutchinsonian natural philosophy," *History of Science, 18* (1980), 1–24.
11 *PW*, p. 132; BM Add MSS 47,525, f96v; *Philosophical Lectures*, pp. 106–7.
12 Robert Fox, *The Caloric Theory of Gases, from Lavoisier to Regnault* (Oxford, 1971).
13 T. Beddoes, "Letter on certain points of history relative to the component parts of the alkalis," Nicholson's *Journal of Natural Philosophy . . . , 21* (1808), 139–41; Davy, *Works*, *8*, 348.
14 BM Add. MSS 47,526, f8v.
15 STC, *Table Talk*, June 29, 1833; CM, E. Swedenborg, *Oeconomia Regni Animalis . . .* (London and Amsterdam, 1740; Amsterdam, 1741), p. 361, copy in Swedenborg Society, London; CM, Nicholson's *Journal, 26* (1810), 31, transcribed by Sara Coleridge in VCL MS BT 37.7.
16 VCL S MS 29, *1*, f74, *2*, f210; *BL*, *1*, 74 (chap. 6).
17 Lavoisier pays tribute to Condillac's *Logique* in the preface to his *Traité élémentarie de chimie*, 2 vols. (Paris, 1789). Coleridge condemns the *Logique* in *TL*, p. 61 n.

18 *BL*, *1*, 85 (chap. 7); VCL S MS 29, *1*, f7; *BL*, *1*, 163 (chap. 12) (cf. *CN*, *1*, 920 [1801]).
19 *CL*, *2*, 708.
20 P. 14. A presentation copy inscribed to Joseph Cottle is in VCL.
21 G. N. G. Orsini, *Coleridge and German Idealism* (Carbondale, Ill., 1969), pp. 131–9.
22 *CL*, *5*, 14. See also *CL*, *5*, 421.
23 W. D. Wetzels, "Aspects of natural science in German romanticism," *Studies in Romanticism*, *10* (1971), 44–59, at 46–7.
24 W. D. Wetzels, "Johann Wilhelm Ritter: Physik im Wirkungsfeld der deutschen Romantik," Ph.D. diss., Princeton University, 1968; J. W. Ritter, *Beweis, das ein beständiger Galvanismus den Lebensprocess in dem Thierreich begleitete* (Weimar, 1798), p. 171.
25 J.-C. Guédon, "The still life of a transition: chemistry in the Encyclopédie," Ph.D. diss., University of Wisconsin, 1974.
26 A. Gode von Aesch, *Natural Science in German Romanticism* (New York, 1941), p. 12.
27 H. A. M. Snelders, "Romanticism and *Naturphilosphie* and the inorganic natural sciences, 1797–1840: an introductory survey," *Studies in Romanticism*, *9* (1970), 193–215, at 197.
28 Aesch, *Natural Science in German Romanticism*, p. 11.
29 Wetzels, "Natural science in German romanticism," p. 53.
30 Aesch, *Natural Science in German Romanticism*, p. 117.
31 *Philosophical Lectures*, p. 176, quoted in T. McFarland, *Coleridge and the Pantheist Tradition* (Oxford, 1969), p. 207. See Introduction to this book.
32 *AR*, p. xi; K. Coburn, *The Self-Conscious Imagination* . . . , Riddell Memorial Lectures, (Oxford, 1974).
33 *PW*, *1*, 113–14.
34 H. C. Oersted, *The Soul in Nature*, trans. L. Horner and J. B. Horner (London, 1852; reprint ed., London, 1966), p. 18.
35 D. M. Knight, "The physical sciences and the romantic movement," *History of Science*, *9* (1970), 54–75, at 57, 72; K. Raine, "Thomas Taylor, Plato and the English Romantic movement," *British Journal of Aesthetics*, *8* (1968), 99–123.
36 Aesch, *Natural Science in German Romanticism*, p. 6; H. A. M. Snelders, "De invloed van Kant, de Romantiek en de 'Naturphilosophie' op de anorganische naturwetenschappen in Duitsland," *Dissertation Abstracts International*, *37*, (1976, 1977), 1558C; R. Calinger, "Kant and Newtonian science: the pre-critical period," *Isis*, *70* (1979), 349–62; G. J.Brittan, Jr., *Kant's Theory of Science* (Princeton, N.J., 1978).
37 The argument in this section is much indebted to B. Gower, "Speculation in physics: the history and practice of *Naturphilosophie*," *Studies in the History and Philosophy of Science*, *3* (1973), 301–56, esp. pt. 1, pp. 301–27.
38 Fichte, *Erste und zweite Einleitung in die Wissenschaftslehre* (Hamburg, 1961), p. 12, trans. by Gower, in "Speculation in physics," p. 307.
39 Orsini, *Coleridge and German Idealism*, p. 175.
40 *CL*, *2*, 673–4 (1801); *BL*, *1*, 101–2 (chap. 9). A copy read by Coleridge is in BM, cat. no. C.126f.13.
41 *CL*, *4*, 792.

42 G. Marcel, *Coleridge et Schelling* (Paris, 1971); Collegium philosophicum Jenense, 1, *Die Philosophie des jungen Schelling: Beiträge zur Schelling-Rezeption in der DDR* (Weimar, 1977).
43 Gower, "Speculation in physics," p. 311.
44 Orsini, *Coleridge and German Idealism*, p. 199.
45 Schelling, *Sämmtliche Werke* (Stuttgart and Augsburg, 1856–61), *3*, 272, trans. by Gower in "Speculation in physics," p.314.
46 Gower, "Speculation in physics," p. 315.
47 H. Zeltner, "Das Identitässystem," in *Schelling: Einführung in seine Philosophie*, ed. H. M. Baumgartner (Freiburg and Munich 1975), pp. 75–94. Schelling discusses his "Gesetz der Identität" in Von der Weltseele (Hamburg, 1809), p. xxiii. See also Orsini, *Coleridge and German Idealism*, p. 202.
48 McFarland, *Pantheist Tradition*, pp. 80–106; F. W. J. von Schelling, *Einleitung zu seinem Entwurf eines Systems der Naturphilosophie* . . . (Jena and Leipzig, 1799), p. 4. A copy of the *Einleitung* bearing STC's notes is in Dr. Williams's Library, London.
49 Schelling, *Einleitung*, pp. 22, 26–33.
50 Quoted by R. C. Stauffer, "Speculation and experiment in the background of Oersted's discovery of electromagnetism," *Isis*, *48* (1957), 33–50, at 36.
51 *Von der Weltseele*, p. vi; Aesch, *Natural Science in German Romanticism*, p. 149.
52 *Von der Weltseele*, p. viii.
53 Ibid., p. 3.
54 I. Kant, *Metaphysische Anfangsgründe der Naturwissenschaft* (Riga, 1786); Gower, "Speculation in physics," p. 320.
55 Schelling, "Allgemeine Deduction." The BM has the following works by Schelling, with STC's marginal notes: *F. W. J. Schelling's philosophische Schriften* (Landshut, 1809), cat. no. C.126.g.7; *Darlegung des wahren Verhältnisser der Naturphilosophie zu der verbesserten Fichte'schen Lehre* (Tübingen, 1806), cat. no. C.126.f.7; *F. W. J. Schelling's Denkmal der Schrift von den göttlichen Dingen &c. des Herrn Friedrich Jacobi* . . . (Tübingen, 1812), cat. no. C.126.f.8(1); *Ideen zu einer Philosophie der Natur* . . . (Landshut, 1803), cat. no. C.43.b.9; *Philosophie und Religion* (Tübingen, 1804), cat. no. C.126.f.8(2); *System des transcendentalen Idealismus* (Tübingen, 1800), cat. no. C.43.b.10; *Ueber die Gottheiten von Samothrace* (Tübingen, 1815), cat. no. C.126.f.8(3). See G. Whalley's forthcoming edition of the marginalia in *CC*. A useful translation is Schelling, *System of Transcendental Idealism*, trans. P. Heath, intro. M. Vater (Charlottesville, Va., 1978).
56 McFarland, *Pantheist Tradition*, pp. 60, 121, 160.
57 *BL*, *1*, 102 (chap. 9). See *Biographia Literaria*, *CC* (forthcoming).
58 STC wanted to read all Schelling's books: *CN*, *3*, 4307.
59 Wetzels, "Natural science in German romanticism," pp. 52–6; Ritter, *Beweis*; C. A. Cullotta, "German biophysics, objective knowledge, and romanticism," *Historical Studies in the Physical Sciences*, *4* (1974), 3–38; K. L. Caneva, "From galvanism to electrodynamics: the transformation of German physics and its social context," *Historical Studies in the Physical Sciences*, *9* (1978), 63–159.

60 CM, H. C. Oersted, *Ansicht der chemischen Naturgesetze* (Berlin, 1812), p. 42, copy in BM, published in K. Coburn, *Inquiring Spirit: A New Presentation of Coleridge* (London, 1951), p. 249.
61 See, e.g., Dulong to Berzelius, Oct. 2, 1820, in *Jac. Berzelius Bref*, ed. H. G. Söderbaum, 6 vols. and 3 supps. (Uppsala, 1912–32), *4*, 17.
62 *CL*, *4*, 757–60.
63 *CL*, *4*, 762.
64 *Lay Sermons*, ed. R. J. White, in *CC* (1972), pp. 14–15; J. M. di Stefano Pappageorge, "Coleridge on hope and history," *Dissertation Abstracts International*, *38* (1977), 3519A; J. T. Miller, Jr., "Ideology and enlightenment: the political and social thought of Samuel Taylor Coleridge," *Dissertation Abstracts International*, *39* (1978), 1775A.
65 *Lay Sermons*, pp. 15–16.
66 *Lay Sermons*, p. 17; *BL*, *1*, 31 (chap. 3).
67 *Lay Sermons*, pp. 23–4. I have not found where Bacon said this.
68 *BL*, *1*, 125 (chap. 10).
69 STC, *Literary Remains*, *1*, 263–4.
70 *Friend*, *1*, 16; *CL*, *1*, 260; STC, *Literary Remains*, *1*, 265.
71 *Friend*, *1*, 500–6.
72 STC, *Table Talk*, March 18, 1832. See also *Philosophical Lectures*, p. 283.
73 *CL*, *3*, 3660; K. Coburn, *Experience into Thought: Perspectives in the Coleridge Notebooks* (Toronto, 1979), pp. 35 ff.; Paracelsus, *Au. Philip. Theoph. Paracelsi Bombast ab Hohenheim, . . . Opera Omnia Medico-Chemico-Chirurgica*, 3 vols. (Geneva 1658) (the copy in the Royal College of Surgeons of England bears two marginal notes by Coleridge).
74 S. Vince, *A Complete System of Astronomy*, 3 vols. (London, 1814–23), *1*, 98–102; R. Small, *An Account of the Astronomical Discoveries of Kepler* (London, 1804).
75 Letters, Conversations and Recollections of S. T. Coleridge, ed. T. Allsop, 2 vols. (London, 1836), *1*, 127; BM Egerton MS 2800 f79.
76 *Table Talk*, Oct. 8, 1830. Kepler showed that the square of a planet's period of revolution about the sun (T^2) was proportional to the cube of the distance (d^3), or, rather, that for any two planets, *A* and *B*,

$$\frac{T_A^2}{T_B^2} = \frac{d_A^3}{d_B^3}$$

77 N29 f19; P. Rossi, *Francis Bacon: From Magic to Science* (London, 1968).
78 *Friend*, *1*, 486–7. See also STC, *On the Constitution of the Church and State*, ed. J. Colmer, in *CC* (1976), p. 37; N29 f62^v, about "Bacon when least himself"; and *Table Talk*, Oct. 8, 1830: "Bacon, when like himself – for no man was ever more inconsistent . . . " Bacon's importance for STC's view of scientific method is discussed in Chap. 4.
79 *CL*, *2*, 1032.
80 W. Whewell, review of Mary Somerville's *On the Connexion of the Physical Sciences* in *Quarterly Review*, *51* (1834), 59–60.
81 Robert Hooke, *Posthumous Works* (London, 1705), pp. 24–6; *Friend*, *1*, 484.
82 *Friend*, *1*, 483.
83 Jean Baptiste de Boyer, Marquis d'Argens, *Kabbalistische Briefe* . . ., 8 pts. in 2 vols. (Danzig, 1773–7), copy in BM, cat. no. C.43.a.2.

84 *CN, 3*, 3270 (1808).
85 *Table Talk*, Oct. 8, 1830.
86 M. Nicholson, *Science and Imagination* (Ithaca, N.Y., 1956), chap. 5. STC discusses Gassendi and atomism in N29 ff107v–106, 105v. Background is given in R. H. Kargon, *Atomism in England from Hariot to Newton* (Oxford, 1966). The quotations are from *TL*, pp. 30–1.
87 *TL*, pp. 30–1, based on Steffens, *Beyträge* p. 37.
88 *AR*, p. 306.
89 *AR*, p. 360; STC, *Letters, Conversations, and Recollections, 1*, 127.
90 STC's Newton was molded in the image of positive science, far closer to Voltaire's Newton in his *Lettres philosophiques* or to D. Brewster's Newton in his *Life of Sir Isaac Newton* (London, 1831) than to the Newton of recent historiography. STC would have had far more sympathy for the Newton of J. E. McGuire and P. M. Rattansi, "Newton and the pipes of Pan," *Notes and Records of the Royal Society of London, 21* (1966), 108–43.
91 *Philosophical Lectures*, pp. 114, 405, n. 3. STC's note in BM Add MSS 47,519, f61v, has a related discussion, as do *CL, 2*, 708, and CM, L. Oken, *Erste Ideen zur Theorie des Lichts, der Finsterniss, der Farben und der Wärme* (Jena, 1808), copy in BM, cat. no. C.44.g.4(1).
92 *Table Talk*, Oct. 8, 1830. For Coleridge on "law," see Chap. 4; and O. Barfield, *What Coleridge Thought* (Oxford, 1972), chap. 10.
93 VCL S MS 29, *1*, f39.
94 *TL*, p. 31.
95 *AR*, p. 44.
96 CM, dated 1829, on the flyleaf of the copy of *The Friend* (1818), vol. *2*, in the Pierpont Morgan Library, cat. no. 49357; *AR*, pp. 363–5.
97 "Monologues of the late Samuel Taylor Coleridge," *Fraser's Magazine, 12* (1835), 619–29 at 626.
98 *Lay Sermons*, pp. 33–4.
99 *Friend, 1*, 61; *BL, 1*, 49 n, 120 (chaps. 3, 10); *CL, 4*, 922, 760, 743. See BM Add MSS 47,525, ff61v–2, for the Anglo-Gallican school of naturalists. See also *CN, 2*, 2598. "Psilosophy" is more correctly rendered as "mere wisdom."
100 Cuvier discusses method in his *Lectures on Comparative Anatomy*, trans. W. Ross, 2 vols. (London, 1802), *1*, xxiv–xxv, and in the introduction to *Le Règne animal distribué d'après son organisation* . . . , 4 vols. (Paris, 1817). See also *Friend, 1*, 475 and n. Cuvier's hostility to German metaphysical science appears, e.g., in his *Rapport historique sur les progrès des sciences naturelles depuis 1789* (Paris, 1810), pp. 9–10.
101 Cuvier, *Lectures on Comparative Anatomy, 1*, 38; CM, front flyleaf of J. F. Blumenbach, *Ueber die natürlichen Verschiedenheiten im Menschengeschlechte* . . . , trans. J. G. Gruber (Leipzig, 1798), copy in BM, cat. no. C.126.c.6. STC's use of Cuvier's works is discussed in Chaps. 4 and 7 of this book. Cuvier was born in Würtemberg and studied at Stuttgart. See also *Friend, 1*, 475.
102 *CL, 4*, 743–4, 788.
103 *CL, 3*, 95.
104 BM Add MSS 47,526, ff9–9v. I am indebted to Kathleen Coburn for the gloss on Abdera.
105 *CN, 2*, 3121.

106 *CL*, *4*, 760, 775, 922, 793. STC combined and encapsulated his views of national characteristics and of the history of science in *Friend*, *1*, 421–2:

GERMANY	*ENGLAND*	*FRANCE*
IDEA, or Law anticipated,	LAW discovered,	THEORY invented,
TOTALITY,	SELECTION,	PARTICULARITY,
DISTINCTNESS.	CLEARNESS.	PALPABILITY.

107 *AR*, p. 328.
108 *BL*, *1*, 103 (chap. 9).
109 *AR*, p. 33 n.
110 BM Add MSS 47,525, ff95–94v.
111 P. 156.
112 Saumarez's note of May 30, 1818 in the Bodleian Library's copy of his *Principles of Physiological and Physical Science* . . . (London, 1812).
113 *CL*, *4*, 664; *Zeitschrift für spekulative Physik*, ed. Schelling, 2 vols. (Jena and Leipzig, 1800–1), copy in BM, cat. no. C.126.e.1(1)(2), with STC's marginalia.

Chapter 4
Coleridge and metascience

1 *CN*, *1*, 1561; cf. *CN*, *1*, 556.
2 *CL*, *1*, 349.
3 *Church and State*, p. 13.
4 STC, Commonplace Book, Huntington Library MS HM 8195 f9.
5 VCL S MS 29, *1*, f17.
6 See *S. T. Coleridge's Treatise on Method as Published in the Encyclopaedia Metropolitana*, ed. A. D. Snyder (London, 1934), and *Coleridge on Logic and Learning*, ed. A. D. Snyder (New Haven, Conn., 1929).
7 *Friend*, *1*, 463.
8 Francis Bacon, *Works*, ed. J. Spedding, R. L. Ellis, and D. D. Heath, 14 vols. (London, 1857–74), *4* (new ed., 1870), 19.
9 *Philosophical Lectures*, p. 223.
10 *Friend*, *1*, 488.
11 H. B. Nisbet, *Goethe and the Scientific Tradition* (London, 1972), describes Goethe's comparable resolution drawn from the same traditions.
12 *TL*, p. 86.
13 STC, *Literary Remains*, *2*, 67–8, 83 (Shakespeare lectures, 1818), 54 ("The Drama Generally").
14 CM, I. Kant, *Vermischte Schriften*, 4 vols. (Halle, 1799; Königsberg, 1807), transcribed in VCL MS BT 21, ff24–5. See I. Kant, *Cosmogony*, trans. W. Hastie, intro. G. J. Whitrow (London and New York, 1970).
15 *CN*, *2*, 2934; K. Coburn, *Inquiring Spirit: A New Presentation of Coleridge* (London, 1951), pp. 244–5; N28 ff2v, 4v.
16 CM, H. C. Oersted, *Ansicht der chemischen Naturgesetze* (Berlin, 1812), pp. 42–3, copy in BM.
17 *CL*, *4*, 807.
18 *CL*, *2*, 709, 714. STC later asked Tom Poole to destroy his letter attacking Newton's *Opticks* – a letter written "in the ebulliency of indistinct Conceptions . . . which if I were to die & it should ever see the *Light* would damn me forever, as a man mad with Presumption." *CL*, *2*, 1014.

19 N21 ½ ff41–41ᵛ.
20 N46 fll.
21 BM Egerton MS 2801 f136.
22 N46 f11ᵛ.
23 *Friend, 1*, 483–4.
24 N35 ff25ᵛ–26.
25 *Church and State*, p. 62.
26 *1*, 92 (chap. 8).
27 BM Egerton MS 2801 f136; N28 f68ᵛ.
28 For Buffon, Linnaeus, and Cuvier, see the *Dictionary of Scientific Biography*, ed. C. C. Gillispie. Supplementary material and context are given in J. Roger, *Les Sciences de la vie dans la pensée française au XVIIIᵉ siècle* (Paris, 1963; 2d ed., 1972). The first issue of *Taxon* for 1976 has valuable information about Linnaeus's classification. See also Henri Daudin, *Études d'histoire des sciences naturelles, vol. 1, De Linné à Jussieu: méthodes de la classification et l'idée de série en botanique et en zoologie*, vol. 2, *Cuvier et Lamarck: les classes zoologiques et l'idée de série animale* (Paris, 1926); P. J. Bowler, "Buffon and Bonnet: theories of generation and the problem of species," *Journal of the History of Biology, 6* (1973), 259–81; James L. Larson, *Reason and Experience: The Representation of Natural Order in the Work of Carl von Linne* (Berkeley, Calif., 1971); P. R. Sloan, "The Buffon-Linnaeus controversy," *Isis, 67* (1967), 356–75.
29 G. Cuvier, *Le Règne animal distribué d'après son organisation* . . . , 4 vols. (Paris, 1817), *1*, xiv.
30 *Friend, 1* 466–70.
31 Ibid., p. 466. There are modern reprints of Linnaeus's key works: *Systema naturae* (1735; reprint ed., Nieuwkoop, Holland, 1964); *Fundamenta Botanica* and *Bibliotheca Botanica* (1736; reprint ed., Munich, 1968); *Philosophia Botanica* (1751; reprint ed., Codicote, Herts., and New York, 1966); *Species Plantarum*, intro. W. T. Stearn (1753; reprint ed. in 2 vols., London, 1957–9).
32 Charles François Brisseau de Mirbel (1776–1854) was a botanist and author of *Elémens de physiologie végétale et de botanique*, 3 vols. (Paris, 1815). STC read his articles in *Quarterly Journal of Science, Literature, and the Arts*, ed. W. T. Brande, *6* (1818), 257–64, *6* (1819), 20–31, 210–26. See N17 ff128–9.
33 *Friend, 1*, 468–9. Thomas Andrew Knight (1759–1838) was an English horticulturist and friend of Humphry Davy. Ellis was probably Daniel Ellis, author of *An Inquiry into the Changes Induced in Atmospheric Air, by the Germination of Seeds, the Vegetation of Plants, and the Respiration of Animals* (Edinburgh, 1807); STC refers to his work in a marginal note to Steffens's *Beyträge*, copy in BM, and elsewhere. Barbara Rooke's editorial notes to the text of *The Friend* identify the other botanists named here by STC. Other criticisms of Linnaeus by STC are in N25 f86ᵛ and N30 f64.
34 N28 f69ᵛ; BM Egerton MS 2801 f136.
35 *Scrofula*, f18.
36 N28 ff72ᵛ–3, N29, f82. The classic account of *The Great Chain of Being* is A. O. Lovejoy's book of that title (Cambridge, Mass., 1936).
37 STC's reservations are close to Cuvier's in *Lectures on Comparative Anatomy*, trans. W. Ross, 2 vols. (London, 1802), pp. xxii ff., and *Le Règne*

animal, *1*, vi ff., 10 ff. STC's discussion of types in N25 ff86ᵛ–8, N27 f71, N28 ff72ᵛ–73, N29 f81, and N56 f9 is surely based on Cuvier's use of "type" as an example of a class – species or genus – best exhibiting the essential characters of that class. There is a helpful account in P. L. Farber, "The type-concept in zoology during the first half of the nineteenth century," *Journal of the History of Biology*, *9* (1976), 93–119.

38 CM, front flyleaf of J. F. Blumenbach, *Ueber die natürlichen Verschiedenheiten im Menschengeschlechte . . .*, trans. J. G. Gruber (Leipzig, 1798), copy in BM.
39 *Friend*, *1*, 470.
40 "Outlines of the history of logic," VCL MS BT 16, f34.
41 *BL*, *1*, 202 (chap. 13).
42 CM, F. W. J. von Schelling, *Ideen zu einer Philosophie der Natur . . .* (Landshut, 1803), pp. 302–3, copy in BM; N30 f49.
43 *Friend*, *1*, 490.
44 E.g., *CN*, *1*, 1831 (Jan. 1804): "In all my metaphysical Speculations never to forget to apply my Theory to Dogs, Cats, Horses, &c ? – 'If it be true, ought not they to be so & so?' – And are they? – an important Preventive of gross Blunders."
45 N34 f5.
46 N28 ff22ᵛ–3.
47 See, e.g., BM Egerton MS 2800 ff155–155ᵛ.
48 N27 ff67ᵛ–8.
49 *Works*, *1*, 148–9, STC's trans.
50 *TL*, p. 30; *Friend*, *1*, 489.
51 *Friend*, *1*, 94 n.
52 Ibid., pp. 493–4 n.
53 N28 f39.
54 F. H. Anderson, *The Philosophy of Francis Bacon* (Chicago, 1948).
55 N29 ff102–96.
56 Nisbet, *Goethe*, pp. 39, 76.
57 *Friend*, *1*, 481.
58 Ibid., pp. 474–5.
59 E. Home, "On the formation of fat in the intestine of the tadpole, and on the use of the yolk in the formation of the embryo in the egg," *Philosophical Transactions*, *106* (1816), 301–10, contained an account (pp. 306–10) of Hatchett's observation. See also *Friend*, *1*, 474–5 n.
60 Cuvier, *Lectures on Comparative Anatomy*, *1*, xxii, mentioned by STC in *CN*, *3*, 4358; *Friend*, *1*, 475 n.
61 NFº f57ᵛ.
62 "On the Divine Ideas," in the Commonplace Book, Huntington Library MS HM 8195.
63 *Friend*, *1*, 467 n; see also *BL*, *1*, 69 n (chap. 5).
64 STC, *Statesman's Manual*, in *Lay Sermons*, *CC*, p. 79.
65 "On the Divine Ideas," Huntington MS HM 8195 f31.
66 Ibid., f33.
67 *AR*, p. 158 n.
68 *Lay Sermons*, p. 70.
69 A. J. Harding, "Development and symbol in the thought of S. T. Cole-

ridge, J. C. Hare, and John Sterling," *Studies in Romanticism, 18* (1979), 29–48; *BL, 1* 100 (chap. 9).

70 *Friend, 1*, 492.

71 *CN, 2*, 2546 (Apr. 14, 1805).

72 *PW*, p. 242, lines 59–62.

73 *PW*, p. 132, lines 19–20. The bulk of the poem was drafted in 1796–7.

74 *AR*, p. xvii.

75 *AR*, p. 229 n.

76 *Statesman's Manual*, p. 30. The OED gives the first use of translucence in this sense – "the action or fact of shining through" – as STC's in 1826.

77 J. H. Frere (1769–1846), fellow of Caius College, Cantab., undersecretary of state in the foreign office, 1799, and member of the Privy Council, 1805, was STC's good friend and patron. The note in the presentation copy is on p. 3 of copy 3 of *Aids to Reflection* (1825), in the Berg Collection, New York Public Library.

78 *AR*, p. 182.

79 Works by Swedenborg bearing notes by STC include *Prodromus philosophiae ratiocinantis de infinito, et causa finali creationis: deque mechanismo operationis animae et corporis* (Dresden and Leipzig, 1734), Huntington Library 132848; *Regnum Animale anatomice, physice et philosophice perlustratum*, 2 pts. in 1 vol. (The Hague, 1744; London, 1745), and *Oeconomia Regni Animalis* . . . (London and Amsterdam, 1740; Amsterdam 1741), copies in Swedenborg Society, London; and *De Equo Albo de quo in Apocalypsi, cap: XIX* . . . (London, 1758), copy in BM, cat. no. C.44.9.5. C. A. Tulk (1786–1849) is not to be confused with Alfred Tulk, translator of Oken's *Elements of Physiophilosophy* (London, 1847). Among STC's letters to C. A. Tulk are *CL, 4*, 767–76, *5*, 17–19. See Raymond H. Deck, Jr., "New light on C. A. Tulk, Blake's nineteenth-century patron," *Studies in Romanticism, 16* (1977), 271–36.

80 Pierpont Morgan Library, MS MA 1853; cf. *variatim CL, 5*, 326 (STC to Tulk, Jan. 26, 1824).

81 *Literary Remains, 2*, 50–1 ("The Drama Generally").

82 *CL, 5*, 19 (Jan. 20, 1820).

83 Swedenborg, *Oeconomia*, pt. 1, p. 368.

84 N26 f66v (May 15, 1826), specifically referring to Swedenborg's *De Coelo et . . . Inferno* (London, 1758).

85 Huntington MS HM 8195 ff7, 13. For this section, see the admirable discussion in O. Barfield, *What Coleridge Thought* (Oxford, 1972), chap. 10.

86 N26 ff119–119v.

87 BM Egerton MS 2801 f139.

88 NQ f53v.

89 Huntington MS HM 8195 f33.

90 *Friend, 1*, 157, 158.

91 CM, Swedenborg, *Oeconomia*, pt. 2, p. 199.

92 BM Egerton MS 2800 f189v. *AR*, pp. 56–8, presents law as an antecedent unity in the world, explaining the parts, and points to a corresponding unity in the material system. STC illustrates this with an account of a crocus.

93 T. McFarland, *Coleridge and the Pantheist Tradition* (Oxford, 1969); VCL S MS 29, *2*, f61.
94 CM, Nicholson's *Journal, 22* (1809), 162, transcribed in VCL MS BT 37.7.
95 *Friend, 1*, 497–8.
96 CM, Richard Hooker, *Of the Lawes of Ecclesiastical Politie* (London, 1682), p. 72, copy in BM.
97 N39 f6[v].
98 VCL S MS F.2.11. Similar statements occur in CM in a copy of *The Friend* (1818), *3*, 252, Pierpont Morgan Library 49356–8 w/18/C; *TL*, p. 25; and *Friend, 1*, 459–61. J.-L. Gay-Lussac's law of combining gas volumes is typical of the laws recognized by the vast majority of scientists in the early nineteenth century; a valuable account is M. Crosland, *Gay-Lussac, Scientist and Bourgeois* (Cambridge, 1978), chap. 3.
99 *Friend, 1*, 511, 458; *BL, 1*, 175 (chap. 12).
100 VCL S MS 29, *1*, f87. In BM Egerton MS 2800 f189[v], STC explains that when the philosopher contemplates cosmogony, "He is in the region of *Ideas*: and his End and his Aim is the reduction of Phaenomena to Ideas, and the construction of a system of *Thoughts* that shall correspond to the System of Things, and render the *sensible* world intelligible."
101 *AR*, pp. 58 n, 61.
102 Kant's *Critique of Teleological Judgement* constitutes part 2 of his *Critik der Urtheilskraft* (Berlin, 1799); a copy of this work bearing CM is in the BM. Immanuel Kant, *The Critique of Judgement*, ed. and trans. J. C. Meredith (Oxford, 1973), has valuable analytical indexes; the index entries "Nature" and "Organism" are especially relevant here. A useful general account is S. Poggi, "Teleologia, spiegazione scientifica, e materialismo dialettico in alcune interpretazioni della Kritik der Urteilskraft," *Rivista di Filosofia, 67* (1976), 497–521.
103 CM, Nicholson's *Journal, 26* (1810), 243, transcribed in VCL MS BT 37.7.
104 Pierpont Morgan MS MA 1854. A similar argument is in *BL, 1*, 86–7 (chap. 7). A copy of I. Kant, *Critik der reinen Vernunft* (Leipzig, 1799), bearing CM is in BM. See also Huntington MS HM 8195 f170.
105 *Friend, 1*, 517–18; *BL, 1*, 163 (chap. 12).
106 *Posterior Analytics*, bk 1., chap. 1, 71b, trans. G. R. G. Mure, in *The Basic Works of Aristotle*, ed. R. McKeon (New York, 1941), p. 111.
107 *Friend, 1*, 464.
108 F1.
109 *Table Talk*, June 29, 1833.
110 *CN, 3*, 4171; M. Crosland, "Lavoisier's theory of acidity," *Isis, 64* (1973), 306–25.
111 Cuvier, *Lectures on Comparative Anatomy, 1*, xxiv–xxv.
112 N28 ff12–13.
113 C. W. Miller, "Coleridge's concept of nature," *Journal of the History of Ideas, 25* (1964), 77–96, raises several of the major issues that I attempt to resolve in this section.
114 H. Steffens, "Recension der neuern naturphilosophischen Schriften des Herausgebers von Dr. Steffens aus Copenhagen," *Zeitschrift für spekulative Physik, 1* (1800), 1–48. Vols. 1–2 of the *Zeitschrift* bearing CM are in BM.

Steffens's works annotated by STC are *Anthropologie*, 2 vols. (Breslau, 1822), copy in Dr. Williams's Library, London; and, all in BM, *Beyträge*; *Caricaturen des Heiligsten*, 2 vols. (Leipzig, 1819–21); *Die Gegenwartige Zeit* (Berlin, 1817); *Geognostisch–geologische Aufsätze* . . . (Hamburg, 1810); *Grundzüge*; *Ueber die Idee der Universitäten* (Berlin, 1809).

115 Steffens, *Grundzüge*, p. 16: "Die Natur braucht, um erkannt zu werden, kein fremdes Prinzip, sondern ist in sich selbst gegründet, d.h. absolut" (Nature needs no foreign [external] principle in order to be known, but is grounded in itself, i.e. absolute). STC in Oct. 1818 refers to Schelling, *Einleitung*, p. 36, for an illustration of "the one πρωτον ψευδος – the making *Nature* absolute." *CN*, *3*, 4449 and n. STC, in a marginal note to the copy of the *Einleitung* now in Dr. Williams' Library (pp. 36–7), gives "an instance of the error [in]to which Schelling is led by making *Nature absolute*." N28 f36[v] describes Schelling's system as a form of "spinozism." STC's comments about Schelling and Catholicism are in *Philosophical Lectures*, pp. 390–1.

116 VCL S MS 29, *1*, ff78–9.

117 P. 11.

118 Steffens, *Beyträge*, p. 256.

119 N56 ff9–9[v]. STC uses "physiogony" widely, e.g., N26 ff20[v], 121[v]–5[v]; N49 ff29–29[v]; N65 ff2–2[v]. J. H. Green used STC's terms in *Vital Dynamics* (London, 1840), p. 101: "The three great divisions into which all natural sciences resolves itself are:

Physiography, or Description of Nature;
Physiology, or Theory of Nature; and, lastly,
Physiogony, or History of Nature."

Richard Owen attended some of Green's lectures, and thence introduced some of STC's terms into comparative anatomy.

120 N26 ff122–3; VCL S MS F2.11; Barfield, *What Coleridge Thought*, p. 132.

121 Barfield, *What Coleridge Thought*, p. 58.

122 *Church and State*, p. 196, cited in Barfield, *What Coleridge Thought*, p. 44.

123 McFarland, *Pantheist Tradition*; L. Metzger, "Coleridge's vindication of Spinoza: an unpublished note," *Journal of the History of Ideas*, *21* (1960), 279–93.

124 *Friend*, *1*, 467 n; BM Egerton MS 2800 f189[v]: "Nature restraining itself, and [unifying] itself; and Nature unfolding itself from within, calling itself forth from without, and [? in every way] *producing itself*."

125 VCL S MS 29, *1* f86.

126 N59 f5.

127 *CN*, *3*, 4449; CM, G. H. Von Schubert, *Allgemeine Naturgeschichte oder Andeutungen zur Geschichte und Physiognomik der Natur* (Erlangen, 1826), p. 629, copy in BM.

128 N29 f118[v].

129 N21 ½ f50[v]; Lovejoy, *Great Chain of Being*.

130 *CL*, *6*, 598, 597 (July 27, 1826).

131 *TL*, p. 41.

132 BM Egerton MS 2801 f158; NF°, f57[v].

133 BM Egerton MS 2801 ff53, 233[v].

134 Ibid., ff143–4.

135 N18 ff166[v]–7.

136 *BL, 1*, 195 (chap. 13).
137 N26 f122; VCL S MS 29, *1*, ff43–4.
138 CM, Hooker, *Lawes of Ecclesiastical Politie*, p. 27. A similar statement is in Green, *Vital Dynamics*, p. 40. There is a useful account on pp. 51–6 of *Vital Dynamics*, App. A, "Evolution of the idea of a power," in which power is presented as the analogue of will, when applied to outward nature as will with the abstraction of intelligence.
139 CM, Nicholson's *Journal, 22* (1809), 162, transcribed in VCL MS BT 37.7.
140 *CL, 4*, 690 (Nov. 10, 1816).
141 CM, L. Oken, *Lehrbuch der Naturgeschichte*, 6 vols. (Leipzig and Jena, 1813; Jena, 1816–26), *1*, 41, copy in BM. See also P. C. Mullen, "The romantic as scientist: Lorenz Oken," *Studies in Romanticism, 16* (1977), 381–99.
142 CM, Behmen [Boehme], *Works, 1*, pt. 1, p. 24, copy in BM. The relations between quality and quantity are discussed later in this chapter and in Chap. 6.
143 VCL S MS 29, *1*, f83 presents STC's threefold division of powers into substantive, modifying, and ordinant.
144 T. Beddoes, "Letter on certain points of history relative to the component parts of the alkalis," Nicholson's *Journal, 21* (1808), 139–41. T. H. Levere, *Affinity and Matter: Elements of Chemical Philosophy, 1800–1865* (Oxford, 1971), chap. 2, discusses Davy's rejection of imponderable fluids. For Franklin, see I. B. Cohen, *Franklin and Newton . . .* (Cambridge, Mass., 1966); and J. L. Heilbron, *Electricity in the Seventeenth and Eighteenth Centuries* (Berkeley and Los Angeles, 1979).
145 VCL S MS 29, *1*, f33; BM Egerton MS 2801 f139v; VCL S MS 29, *1*, f87.
146 *BL, 1*, 196 (chap. 13); B. Gower, "Speculation in physics: the history and practice of *Naturphilosophie*," *Studies in the History and Philosophy of Science, 3* (1973), 301–56, at 324–7; Schelling *Von der Weltseele*, p. 3.
147 *CL, 4*, 883 [Nov. 24, 1818].
148 CM, Behmen, [Boehme], *Works, 1*, pt. 1, pp 22/23–4, *2*, pt. 1, pp. [1]–4; T. McFarland, "Coleridge and Boehme," in *Pantheist Tradition*, pp. 325–32.
149 *Friend, 1*, 94 n; BM Egerton MS 2801 f15v; *BL, 1*, 103 (chap. 9); Barfield, *What Coleridge Thought*, pp. 179–93.
150 *CL, 4*, 790 (Dec. 12, 1817).
151 BM Egerton MS 2800 f39.
152 *TL*, p. 69; *CL, 4*, 771 (Sept. 1817).
153 CM, Runge, *Neueste Phytochemische Entdeckungen zur Begründung einer wissenschaftlichen Phytochemie*, 2 pts. (Berlin, 1820–1), p. 44, copy in BM.
154 *Friend, 1*, 94 n.
155 Gower, "Speculation in physics," p. 326; CM, Behmen, [Boehme], *Works, 1*, pt. 1, pp. 22/23–4.
156 *Friend, 1*, 94 n.
157 *Church and State*, p. 233; CM, front flyleaf of Swedenborg, *Prodromus*; BM Egerton MS 2801 f125; VCL S MS 29, *1*, f15.
158 BM Egerton MS 2801 ff129v, 139v.
159 *Friend, 1*, 478–81.
160 Ibid., p. 94 n; *Church and State*, passim.

161 *Table Talk*, Apr. 5, 1832.

162 CM Goldfuss, *Handbuch der Zoologie*, 2 vols. (Nürnberg, 1820), *1*, +3, copy in VCL. (CM fly pagination is explained in the List of Abbreviations at the head of the notes.)

163 CM, *Quarterly Journal of Foreign Medicine and Surgery, 1* (1818–19), transcribed by Sara Coleridge in VCL MS BT 37.

164 *CN*, *3*, 4226 ([1814–]15); BM Egerton MS 2800 f167 (Aug. 1817); *Church and State*, p. 233 (letter of 1830 or later).

165 N26 f99^v.

166 *De generis humani varietate nativa* (Göttingen, 1775; 3d ed., Göttingen, 1795). STC's annotated copy of Blumenbach, *Ueber die natürlichen Verschiedenheiten*, is in BM. There is a convenient English translation, *The Anthropological Treatises of Johann Friedrich Blumenbach . . . and the Inaugural Dissertation of John Hunter, M.D., on the Varieties of Man*, trans. and ed. T. Bendyshe (London, 1865). See also J. H. Haeger, "Coleridge's speculations on race," *Studies in Romanticism, 13* (1974), 333–57; and T. H. Levere, "S. T. Coleridge and the human sciences: anthropology, phrenology, and mesmerism," in *Science, Pseudoscience, and Society*, ed. M. P. Hanen, M. J. Osler, and R. G. Weyant (Waterloo, Ont., 1980), pp. 171–92.

167 (London, 1813), p. 233. There is a recent reprint, intro. and ed. G. W. Stocking (Chicago, 1973). See also H. Odom's article on Prichard in the *Dictionary of Scientific Biography*, ed. C. C. Gillispie.

168 N28 f56; Oken, *Naturgeschichte*, *6*, 1233–4.

169 N26 f99^v. STC annotated two editions of I. Kant, *Anthropologie*: Königsberg, 1798, copy now lost; and Königsberg, 1800, copy in BM. A recent translation is *Anthropology from a Pragmatic Point of View*, trans. V. L. Dowdell (Carbondale, Ill., and London, 1978).

170 CM, *Quarterly Journal of Foreign Medicine and Surgery, 1* (1818–19), VCL MS BT 37; *Table Talk*, Apr. 24, 1832; NF° f16^v; N21 ½ ff53–4; NF° f54.

171 BM Egerton MS 2800 f167.

172 N27 f52^v, written in 1819 – "Pan" is a pun on "all," "totality," and "pentad."

173 N23 ff28–32^v.

174 Derived from N27 f39.

175 N29 f56.

176 *CN*, *3*, 4223 n.

177 Schelling's construction is in his "Allgemeine Deduction," pt. 2, pp. 1–87, where STC would have encountered it. The volume bearing his annotations is in BM. See Chap. 5 of this book.

178 N27 f52^v.

179 *Beyträge*, pp. 262–3; CM, ibid.

180 *Beyträge*; CM, ibid., p. +11.

181 CM, Thomas Browne, *Pseudodoxia Epidemica . . .* , 3d ed. (London, 1658), pt. 1, pp. 58–9, copy in Berg Collection, New York Public Library.

182 N27 f69^v.

183 *CN*, *3*, 4414.

184 Copy in BM. The symbols are also tabulated in "Dialogues concerning the ends and method of Philosophy," N27 f49, and in *Opus Maximum*, VCL S MS 29, *1*, ff193–4. A similar set of correspondences is set out in

STC's MS entitled "On the polar logic," Cornell University Library, Wordsworth Collection.

185 N56 ff9–9v.

186 *CN*, *3*, 4265.

187 *CN*, *3*, 4226.

188 J. Neubauer, "Dr. John Brown (1735–88) and early German Romanticism," *Journal of the History of Ideas*, *28* (1967), 367–82.

189 In *CN*, *3*, 4418, the two lines and their intersection are conceived as thesis, antithesis, and synthesis, respectively.

190 VCL S MS 29, *2*, ff37v–8v. STC refers to Richard Baxter's autobiography, *Reliquiae Baxterianae* . . . (London, 1696), pt. 3, p. 69.

191 N23 ff28–32v.

192 A copy of this work, annotated by STC, is in BM.

193 See n. 177 to this chapter.

194 N23 ff31–2v.

Chapter 5
The construction of the world

1 N26 f1 (1826).

2 N29 f117 (?1823).

3 A. Thackray, *Atoms and Powers* (Cambridge, Mass., 1970); P. M. Heimann and J. E. McGuire, "Newtonian forces and Lockean powers: concepts of matter in eighteenth-century thought," *Historical Studies in the Physical Sciences*, *3* (1971), 233–306.

4 Trans. W. Hastie, intro. M. K. Munitz (Ann Arbor, 1969). STC. read the work in Kant's *Vermischte Schriften*, 4 vols. (Halle, 1799), *1*, 183–520, in copies in BM and University College, London. See also R. Calinger, "Kant and Newtonian science: the pre-critical period," *Isis*, *70* (1979), 349–62.

5 Trans. N. K. Smith (Oxford, 1971), pp. 402–9, A434/B462–A443/B471.

6 The copy annotated by STC is in BM. Quotations are from the English translation by B. Bax, *Metaphysical Foundations of Science* (London, 1883), pp. 147, 150, 170, 172, 191. There is a more recent translation by J. Ellington, *Metaphysical Foundations of Natural Science* (Indianapolis and New York, 1970). See also G. J. Brittan, Jr., *Kant's Theory of Science* (Princeton, N.J., 1978).

7 CM, ca. 1819, Kant, *Metaphysische Anfangsgründe*, p. 80, front flyleaf, pp. 36, 103–4.

8 *Schelling: Ausgewählte Werke*, *4* (Darmstadt, 1967), p. 321.

9 CM, Nicholson's *Journal*, *26* (1810) 131, transcribed in VCL MS BT 37.7.

10 "Allgemeine Deduktion." The copy annotated by STC is in the BM.

11 Ibid., pt. 1, pp. 100–25; Brugmans, *Tentamini philosophica de Materia Medica* (Franeker, 1765), pp. 12–24; *TL*, pp. 87–90.

12 Schelling, "Allgemeine Deduktion," pt. 1, pp. 124–5.

13 Ibid., pt. 2, pp. 6–42.

14 CM, ibid., p. 16.

15 Pp. 22 ff., 40–8.

16 *CN*, *3*, 4244 (1815).

17 *CL*, *4*, 873–6 (Sept. 30, 1818).

18 *CN*, *3*, 4449 (Oct. 1818).
19 E.g., in *TL*, pp. 52–7; *CL*, *4*, 768–9; N23 f32^{v}.
20 CM, F. F. Runge, *Neueste Phytochemische Entdeckungen zur Begründung einer wissenschaftlichen Phytochemie*, 2 pts. (Berlin, 1820–1), p. 46, copy in BM.
21 *Euclid's Elements*, ed. I. Todhunter (London and New York, 1933).
22 *BL*, *1*, 171 (chap. 12).
23 *Opus Maximum*, VCL S MS 29, *2*, f210.
24 *CL*, *6*, 597; BM Egerton MS 2801 ff90, 125 (see also N18 ff167–8); Schelling, *Einleitung*, pp. 30–2. The notion of a line fixed at one end but still infinite had been demolished by Aristotle, because there could be no infinite magnitude. *Physics*, bk. 3, chap. 7, 207B. See also *Metaphysics*, bk. 11, chap. 2, 1060B.
25 N23 ff28–32^{v}.
26 VCL S MS 29, *1*, ff59^{v}–61^{v}. See also N27 f55^{v}.
27 CM, Marcus, "Versuch eines Theorie der Entzündung," *Jahrbücher der Medicin als Wissenchaft*, *3*, pt. 1, p. 51, copy in BM.
28 N21 ½ f59^{v} (?1819).
29 N28 ff2^{v}–3.
30 NF° ff69^{v}–70. The third through sixth parts followed on ff70–4^{v}, but they go well beyond the scope of this book.
31 J. R. de J. Jackson's useful phrase.
32 "Outlines of the history of logic," VCL MS BT 16, f24.
33 M. H. Abrams, "Coleridge's 'A light in sound': science metascience, and poetic imagination," *Proceedings of the American Philosophical Society*, *116* (1972), 458–76. Allen Debus, *The Chemical Philosophy: Paracelsian Science and Medicine in the Sixteenth and Seventeenth Centuries*, 2 vols. (New York, 1977), provides an account of the Hermetic and Paracelsian traditions to which Boehme was heir. See also A. Koyré, *La Philosophie de Jacob Boehme* (Paris, 1929); and T. McFarland, *Coleridge and the Pantheist Tradition*, (Oxford, 1969) excursus n. 19, "Coleridge and Boehme." Additional illumination is provided by F. Yates, *The Rosicrucian Enlightenment* (London, 1972), and *Giordano Bruno and the Hermetic Tradition* (London, 1964).
34 *CN*, *3*, 4418 (Aug. 1818).
35 CM, G. A. Goldfuss, *Handbuch der Zoologie*, 2 vols. (Nürnberg, 1820), *1* –3, copy in VCL; *CN*, *4*, 805 (Jan. 12, 1818); BM Egerton MS 2801 f123 (cf. Steffens, *Beyträge*, p. 208); *TL*, pp. 67–8; *CN*, *3*, 4418.
36 N29 ff118^{v}–117^{v}.
37 BM Egerton MS 2800 f189^{v}, written in 1825 or later, but compatible with notebook entries of 1819.
38 *CN*, *3*, 4418, ff14^{v}–15; VCL S MS 29, *1*, f15.
39 Pp. 22–41, at 38.
40 M. J. Petry, *Hegel's Philosophy of Nature*, 3 vols. (London and New York, 1970), *1*, 285, gives a brief account of Eschenmayer. See "Cosmology and astromy" in this chapter for STC's use of his *Psychologie*.
41 VCL S MS 29, *1*, f87; CM, J. G. Eichhorn, *Einleitung in die Apokryphischen Schriften des Alten Testaments* (Leipzig, 1795), pp. 4–5, copy in BM.
42 VCL S MS 29, *1*, ff1, 64–5, 69. ♓ means "opposite to."
43 *CN*, *3*, 3660; *Table Talk*, Sept. 1, 1832.
44 VCL S MS 29, *1*, ff69–70.

45 N27 f49; CM, Emanuel Swedenborg, *De Equo Albo de quo in Apocalypsi, cap: XIX* . . . (London, 1758), copy in BM.
46 CM, H. Steffens, *Geognostisch–geologische Aufsätze* . . . (Hamburg, 1810), pp. 190–259, 313. The context of biblical criticism is discussed by E. S. Shaffer in *"Kubla Khan" and "The Fall of Jerusalem,"* (Cambridge, 1975).
47 *CN*, *3*, 4418, ff15v–16 (Aug. 1818).
48 *CL*, *5*, 18 (Jan. 1820). Cf. Genesis 1: 10–19.
49 E.g., N26 f30 (?1826); N18 f164.
50 CM, J. G. Eichhorn, *Einleitung ins Alte Testament*, 2d ed., 3 vols. in 3 pts. (Leipzig, 1787), *2*, 339–40.
51 N26 ff24–24v; W. Yourgrau and A. D. Breck, eds., *Cosmology, History, and Theology* (New York, 1977).
52 N27 ff39–54.
53 N28 ff53v, 34v.
54 N42 ff32v–3.
55 N42 ff17v–18.
56 N45 f2v. Perhaps STC knew Steffens's splendid *Vollständiges Handbuch der Oryktognosie*, 4 vols. (Halle, 1811–24). For the background of the debate between biblical and scientific interpretation of the age of the earth, see C. C. Gillispie, *Genesis and Geology* (Cambridge, Mass., 1951).
57 *BL*, *1*, 202 (chap. 13).
58 *CN*, *3*, 4319 (Dec. 1816). See also VCL S MS 29, *1*, ff52–3.
59 VCL S MS 29, *1*, f35.
60 E.g., *CN*, *3*, 4226.
61 N27 f56v.
62 CM, Behmen [Boehme], *Works*, *1*, loose leaf recto, copy in BM. In a copy now in BM of G. H. von Schubert, *Allgemeine Naturgeschichte oder Andeutungen zur Geschichte und Physiognomik der Natur* (Erlangen, 1826), p. +1, STC wrote: "Next to that, to which there is no Near, the γυιλτ and the avenging Daemon of my Life, I must place the neglect of Mathematics under the strongest motives, and the most favorable helps and opportunities for acquiring them. Not a week passes in which I do not regret this . . . Oversight of my Youth with a sort of . . . remorse that turns it to a Sin."
63 N46 f5.
64 E. Darwin, *The Botanic Garden, a Poem in Two Parts: Part 1, Containing the Economy of Vegetation; Part 2, The Loves of the Plants, with Philosophical Notes* (pt. 1, London, 1791; pt. 2, Lichfield, 1798), *1*, 67 n; CM, Schubert, *Ansichten* . . . (Dresden, 1808), pp. 126–7, copy in BM. W. Herschel's papers on these subjects include *Philosophical Transactions*, *75* (1785), 40, 213, 77 (1787), 4, *79* (1789), 151, 212, *93* (1803), 339, *98* (1808), 145, *102* (1812), 115, 229.
65 N28 ff88v–9 (1819–20), where STC refers to Long.
66 N40 f11v.
67 W. R. Hamilton, *On a General Method of Expressing the Paths of Light, and of the Planets, by the Coefficients of a Characteristic Function* (Dublin, 1833), presentation copy to STC in BM.
68 E.g., N46 f13 and VCL MS BT 21 f20.
69 NF° f69v.

70 Newton's account is in *Principia*, bk. 3, lemma 4 ff. A good brief account of the relevant history of theories about comets is Petry, *Hegel's Philosophy of Nature*, *1*, 366–7. STC's interest is discussed in *CN*, *3*, 3798 and n. The marginalia on the letter to Bode were to Nicholson's *Journal*, *22* (1809), 210, transcribed in VCL MS BT 37.7.

71 *CN*, *3*, 3840; *CL*, *3*, 414.

72 BM Egerton MS 2801 f71; *CL*, *4*, 954–6; CM, Stanley, *History of Philosophy* (London, 1701), p. 405, copy in Berg Collection, New York Public Library.

73 CM, Nehemiah Grew, *Cosmologia Sacra* (London, 1701), pp. 8–10, copy in BM; G. L. L. Buffon, *Histoire naturelle*, 44 vols. (Paris 1749–1804), supp. vol. *5*, *Epoques de la nature*.

74 VCL S MS 29, *1*, f4. See also CM, Steffens, *Geognostisch–geologische Aufsätze*, pp. 219–20; and CM Behmen [Boehme], *Works*, *1*, pt. 1, p. 250.

75 N40 f11v; CM, Kant *Vermischte Schriften*, *1*, 353–4, transcribed in VCL BT 21 f24; N21 ½ f49v.

76 *CN*, *3*, 3802 (1810); N3 ½ f82v (ca. 1823). In *CN*, *3*, 4333 (1817), STC stated that "the whole Planet has been actuated by a Planetary Life." See also N26 f28 (ca. 1826).

77 Schubert, *Ansichten*, pp. 104, 105; CM, H. Boerhaave, *A New Method of Chemistry*, trans. P. Shaw and E. Chambers, 2 pts. in 1 vol. (London, 1727), pt. 1, p. 120, transcribed in VCL MS BT 37.3.

78 C. A. von Eschenmayer, *Psychologie* (Stuttgart and Tübingen, 1817), p. 151. The copy with CM is in BM.

79 CM, Steffens, *Geognostisch–geologische Aufsätze*, p. 313; Steffens, *Grundzüge*, pp. 130, 146.

80 VCL S MS 29, *1*, ff34–5; NF° f69v. In *Table Talk*, Mar. 18, 1832, STC said: "There is no doubt but that astrology of some sort or other would be the last achievement of astronomy: there must be chemical relations between the planets; the difference of their magnitudes compared with that of their distances is not explicable otherwise."

81 N28 f85v; Steffens, *Geognostisch–geologische Aufsätze*, p. 269; BM Egerton MS 2801 f15v.

82 N26 f24; CM, Schelling, *Einleitung*, p. −3, copy in Dr. Williams's Library, London. STC's references are to Laplace's *Mécanique céleste*, 5 vols. (Paris, 1799–1825), a brilliant work of mathematical analysis applied to celestial motions, and entirely beyond STC's reach, except through nontechnical reviews. He could but seems not to have read the *Système du monde*, in which the cosmology and astronomy are presented without the supporting mathematical apparatus. For Le Sage, see the *Dictionary of Scientific Biography*. See also CM, Schelling, *Einleitung*, pp. 6, +2, 102–3.

83 Table Talk, Oct. 8, 1830; Max Caspar, *Kepler*, trans. and ed. Doris Hellman (London and New York, 1959). In STC's own day, Kepler's reputation in England was enhanced by Robert Small, *An Account of the Astronomical Discoveries of Kepler* (London, 1804).

84 N28 ff16–18v, 85v; Steffens, *Grundzüge*, pp. 29–35. The quotation is from *Grundzüge*, pp. 34–5, in a translation that will be published in *CN*, *4*.

85 *Geognostisch–geologische Aufsätze*, pp. 267–73.

86 N29 f22v.
87 See also *CN, 3*, 4435, 4436, and nn.
88 Eschenmayer, *Psychologie*, pp. 453, 457, 488–90, 530 ff.; CM, ibid., pp. 488–90; N29 f23.
89 CM, *Allgemeine Naturgeschichte*, front flyleaf, copy in BM.
90 A. Volta, "On the electricity excited by the mere contact of conducting surfaces of different kinds," *Philosophical Transactions, 90* (1800), 403–31.
91 Robert Fox, *The Caloric Theory of Gases, from Lavoiser to Regnault* (Oxford, 1971).
92 N28 ff23v–4; W. T. Brande, *A Manual of Chemistry* (London, 1819).
93 H. Guerlac, "Chemistry as a branch of physics: Laplace's collaboration with Lavoisier," *Historical Studies in the Physical Sciences*, 7 (1976), 193–276; T. H. Levere, "The interaction of ideas and instruments . . .," in G. L'E. Turner and T. H. Levere, *Van Marum's Scientific Instruments in Teyler's Museum*, vol. 4 of *Martinus van Marum: Life and Work*, ed. E. Lefebvre and J. G. de Bruijn (Leyden, 1973), pp. 103–22; D. Arnold, "The Mecánique Physique of . . . Poisson: the evolution and isolation in France of his approach to physical theory (1800–1840)," Ph.D. diss., University of Toronto, 1978. There is an extensive account of caloric and its role in forming gases in A.-L. Lavoisier, *Traité élémentaire de chimie*, 2 vols. (Paris, 1789), trans. R. Kerr as *Elements of Chemistry* (Edinburgh, 1790), pt. 1, "Of the formation and decomposition of aeriform fluids, – of the combustion of simple bodies, and the formation of acids." See also R. J. Morris, "Lavoisier and the caloric theory," *British Journal for the History of Science, 6* (1972), 1–38; and J. R. Partington and D. McKie, "Historical studies on the phlogiston theory," *Annals of Science, 2* (1937), 361–404, *3* (1938), 1–58, 337–71, *4* (1939), 113–49.
94 Watt to Darwin, in James P. Muirhead, *The Origin and Progress of the Mechanical Inventions of James Watt*, 2 vols. (London, 1854), 2, 123; Davy, *Works, 2*, 41; W. Herschel, "Experiments on the refrangibility of the invisible rays of the sun," *Philosophical Transactions, 90* (1800), 284–92; Newton, *Opticks*, 4th ed. (London, 1730), queries 1, 4, 5, 7, 8, 9.
95 Schelling, *Ideen*, p. 416; CM, ibid., 406, copy in BM; *Table Talk*, Aug. 8, 1831.
96 N26 ff23–23v (ca. 1826); N28 ff8v–9 (ca. 1819).
97 *Grundzüge*, p. 48; *CL, 4*, 773; N29 f22; BM Egerton MS 2800 f155. See also CM, Behmen [Boehme], *Works, 1*, pt. 1, pp. 249–51; and CM, Goldfuss, *Handbuch der Zoologie, 1*, – 3– – 1.
98 *Opticks*, bk. 1, pt. 1, prop. 1, theor. 1: "Lights which differ in Colour, differ also in Degrees of Refrangibility"; ibid., prop. 2, theor. 2: "The Light of the Sun consists of Rays differently Refrangible"; ibid., prop. 3, theor. 3: "The Sun's Light consists of Rays differing in Reflexibility, and those Rays are more reflexible than others which are more refrangible."
99 *On the Soul*, bk. 2, chap. 7, 4186, trans. J. A. Smith, in *The Basic Works of Aristotle*, ed. R. McKeon (New York, 1941), p. 568.
100 (Reprint ed., New York, 1952), pp. 113–21.
101 *CL*, 2, 709–10, 1046; *Philosophical Lectures*, p. 114, n 3; N21 ½ f61v (?1820); CM, L. Oken, *Erste Ideen zur Theorie des Lichts, der Finsterniss, der Farben und der Wärme* (Jena, 1808), p. 14, copy in BM; CM, Emanuel Swedenborg, *Oeconomia Regni Animalis* . . . (London and Amsterdam,

1740; Amsterdam, 1741), pt. 2, §§257, 199, copy in Swedenborg Society, London.

102 N21 ½ f54ᵛ.

103 *CL*, *3*, (Dec. 7, 1812); N49 f4 (Nov. 1830). STC may have encountered Goethe's work on color in his *Zur Naturwissenschaft Ueberhaupt* (Stuttgart and Tübingen, 1817). In VCL MS LT 47, STC defends Goethe against the *Quarterly Review* of 1814, in terms that indicate familiarity with Goethe's line of thought, but do not suggest intimate knowledge of *Zur Farbenlehre*.

104 *Farbenlehre*, paras. 198, 204, 214; Goethe, *Theory of Colours*, trans. C. L. Eastlake (London, 1840; reprint ed., Cambridge, Mass., 1970), p. xxxvii.

105 J. Dollond, "An account of some experiments concerning the different refrangibility of light," *Philosophical Transactions*, *50* (1758), 733–43.

106 Petry, *Hegel's Philosophy of Nature*, *2*, 359–60; N21 ½ f53; CM, Oken, *Theorie des Lichts*, p. 14; *CL*, *4*, 750–1.

107 Goethe, *Theory of Colours*, pp. 223 ff.; Petry, *Hegel's Philosophy of Nature*, *2*, 146. It is noteworthy that STC gave little attention to Hegel, e.g., reading and annotating less than 100 pp. of vol. *1*, pt. 1, of G. W. F. Hegel, *Wissenschaft der Logik*, 3 pts. in 2 vols. (Nürnberg, 1812–16), copy in BM, cat. no. C.43.a.13.

108 *CN*, *3*, 3606 and n (?1809).

109 CM, Oken, *Theorie des Lichts*, pp. 14, 40.

110 Pp. 47–8. Steffens, *Schriften . . . alt und neu*, 2 vols. in 1 (Breslau, 1821), *2*, 5–35, discusses the significance of color in nature.

111 N49 ff3ᵛ–5ᵛ; BM Egerton MS 2801 f152; Goethe, *Theory of Colours*, pp. 203, 486.

112 N29 ff118–117ᵛ; *CL*, *4*, 750–1; M. H. Abrams, "Coleridge's 'A light in sound': science, metascience, and poetic imagination," *Proceedings of the American Philosophical Society*, *116* (1972), 458–76.

113 N38 f29ᵛ.

114 CM, Swedenborg, *Oeconomia Regni Animalis*, pt. 2, p. 199.

115 NF° f54.

116 Oken, *Theorie des Lichts*, pp. 20, 21, 23, 40; CM, ibid., pp. 14, 40.

117 *CN*, *2*, 3116 (1807); Newton, *Opticks*, bk. 2, fig. 3, and bk. 1, prop. 6, prob. 2. Newton's rings are the concentric colored circles observed when a convex lens of slight curvature is pressed against a sheet of glass, and light is transmitted through them both.

118 VCL MS LT 1. N21 ½ ff53–4ᵛ is an extended discussion of the problem of light and colors, prompted by a consideration of Oken's work. The pentad is in BM Egerton MS 2800 f178ᵛ, watermark 1825; and *Table Talk*, Apr. 24 1832, which also presents a heptad.

119 N27 f63ᵛ; N21 ½ f57ᵛ.

120 NF° ff16–16ᵛ, 54 (1826); CM, July 6, 1830, Oken, *Lehrbuch der Naturgeschichte*, 6 vols. (Leipzig and Jena, 1813; Jena, 1816–26), *1*, +3, copy in BM.

121 John Hunter, *A Treatise on the Blood, Inflammation and Gunshot Wounds, . . . to Which Is Prefixed a Short account of the Author's Life, by His Brother-in-law Everard Home* (London, 1794), p. xxxix. STC refers to this in CM, Behmen [Boehme], *Works*, *1*, pt. 1, pp. 249–51; and CM, Oken, *Naturgeschichte*, *6*, 1234. See also N28 f56.

122 "The Eolian Harp," *PW, 1,* 101, lines 26–9. Abrams, "Coleridge's 'A light in sound,'" is an admirable study of issues discussed in this section, which is therefore brief.
123 *CL, 4,* 750–1. See also STC to Tulk, *CL, 4,* 771, 773 (Sept. 1817); and CM, Oken, *Naturgeschichte,* 1, +1–+3.
124 Young, "Outlines of experiments and inquiries respecting sound and light," *Philosophical Transactions, 90* (1800), 106–50.
125 N29 ff118^v–117^v.
126 Abrams, "Coleridge's 'A light in sound,'" p. 468.
127 Behmen [Boehme], *Works, 1,* pt. 1, p. 43; CM, ibid.

Chapter 6
Geology and chemistry

1 N51 f22.
2 *CL, 1,* 177. John Playfair, *Illustrations of the Huttonian Theory of the Earth* (Edinburgh, 1802; reprint ed., New York, 1956), is the best statement of Hutton's theory.
3 *CN, 1,* 9, 10, 37; Darwin, *The Botanic Garden, a poem in two parts: Part 1, Containing the Economy of Vegetation* (London, 1791), n. xvii.
4 The debate is decribed in R. Porter, *The Making of Geology: Earth Science in Britain, 1600–1815* (Cambridge, 1977); M. J. S. Rudwick, *The Meaning of Fossils: Episodes in the History of Palaeontology* (1951; 2d ed., New York, 1976); and P. J. Bowler, *Fossils and Progress: Palaeontology and the Idea of Progressive Evolution in the Nineteenth Century* (New York, 1976). J. Stock, *Memoirs of the Life of Thomas Beddoes, M.D. . . .* (London and Bristol, 1811), app. 6, publishes part of the Darwin–Beddoes correspondence. Beddoes is inscribed in the medical matriculation records of Edinburgh University for 1780–2 and 1784–86 inclusive. *Philosophical Transactions, 81* (1791), 48–70, and Bodleian Library MS Dep. C. 134 are respectively Huttonian and Wernerian. T. Beddoes, *Four lectures on the Natural History of the Earth* (Oxford, [1790]), and *Lectures on Select Points of Physical Geography* (Oxford, [1790 or 1791]), still lean toward Werner.
5 R. Cardinal, "Werner, Novalis, and the signature of stones," in *Deutung und Bedeutung: Studies in German and Comparative Literature Presented to Karl-Werner Maurer,* ed. B. Schludermann, V. G. Doerksen, R. J. Glendinning, and E. S. Firchow (The Hague and Paris, 1973), pp. 118–32.
6 Werner, *Von den ausserlichen Kennzeichen der Fossilien* (Leipzig, 1774), trans. T. Weaver (Dublin, 1805). See N24 f60 (ca. 1819).
7 A. Ospovat, "Werner, Abraham Gottlob," *Dictionary of Scientific Biography*; Werner, *Short Classification and Description of the Various Rocks,* trans., intro., and ed. A. Ospovat, with facsimile of original (1786) text (New York, 1971); Werner, *On the External Characters of Minerals,* trans. A. V. Carozzi (Urbana, Ill., 1972).
8 Playfair, *Huttonian Theory of the Earth*; V. A. Eyles, "Hutton, James," *Dictionary of Scientific Biography.*
9 Rudwick, *The Meaning of Fossils*; Bowler, *Fossils and Progress.*
10 M. Berman, *Social Change and Scientific Organization: The Royal Institution, 1799–1844* (Ithaca, N.Y., 1978), p. 62; R. Siegfried and R. H. Dott, Jr., "Humphry Davy as geologist, 1805–1829," *British Journal for the History of*

Science, 9 (1976), 219–27; Davy, *Works, 8,* 180–238; W. T. Brande, *Outlines of Geology* (London, 1817), and *A Manual of Chemistry* (London, 1819). STC's notes on the *Manual* are mostly in N27.

11 L. Oken, *Lehrbuch der Naturgeschichte,* 6 vols. (Leipzig and Jena, 1813; Jena 1816–26), *1,* 1.
12 *TL,* p. 40; *CL, 6,* 599; *TL,* p. 42.
13 *TL,* pp. 47–8.
14 *CN, 3,* 4432 (1818).
15 *TL,* pp. 67, 69. See also N27 f74[v]. Steffens's view is given in *Beyträge,* p. 104, and *Geognostisch–geologische Aufsätze . . .* (Hamburg, 1810), pp. 200–1, 228.
16 *TL,* pp. 70–1.
17 *TL* draws heavily on *Beyträge,* pp. 37–8, 273, 275–317.
18 See also Steffens, *Vollständiges Handbuch der Oryktognosie,* 4 vols. (Halle, 1811–24).
19 *Opus Maximum,* VCL S MS 29, *1,* p. 79. A later, more critical, view is expressed in *CL, 5,* 370–2 [1824].
20 VCL MS LT 50 f.
21 BM Egerton MS 2801 f62.
22 CM, Behmen [Boehme], *Works, 2,* pt. 1, pp. 1–4, copy in BM; *CL, 4,* 804 (Jan. 1818).
23 Berman, *Social Change.*
24 N28 ff2[v]–3. Thomas Thomson (1753–1852) was a chemist and a prolific author and editor. Charles Hatchett (1765–1847) was a chemist and W. T. Brande's father-in-law.
25 N29 f1 refers to John MacCulloch, who published chemico-geological papers in the *Quarterly Journal of Science, 10* (1821), 29–51, and *19* (1825), 28–44.
26 Beddoes, "Observations on the affinity between basalts and granite," *Philosophical Transactions, 81* (1791), 48–70, at 69–70 n. Brande, *Outlines of Geology,* pp. 68–9, republished in *Manual of Chemistry,* p. 504, discusses de Saussure, *Voyage, 1,* 166–7. The debacle is described in *Voyage, 1,* 159–60.
27 N27 f55[v]; *Opus Maximum,* VCL S MS 29, *1,* ff59[v]–60[v].
28 Brande, *Manual of Chemistry,* p. 480, and *Outlines of Geology,* p. 79; N28 ff11–12.
29 Braconnot, "On the conversion of ligneous matter into gum, sugar, a particular acid, and ulmin," *Quarterly Journal of Science, Literature, and the Arts, 8* (1820), 386–93; N29 f22; Steffens, *Beyträge,* pp. 168, 170, 174.
30 N29 f23.
31 N29 ff109–108[v].
32 Robinson, *On Books . . .* (London, 1938), p. 307, reprinted in R. W. Armour and R. F. Howe, *Coleridge the Talker* (London, [1840]), p. 333.
33 N20 f42; N26 ff30, 101–2. Cuvier, *Essay on the Theory of the Earth, with Mineralogical Notes and an Account of Cuvier's Discoveries by Prof. Jameson,* trans. R. Kerr, 2d ed. (Edinburgh, 1815), is a likely source for these views, and the copresence of Cuvier and Jameson illustrates the potential conformity of catastrophism with Neptunism. See also VCL MS LT 50f.
34 G. P. Scrope, review of C. Lyell, *Principles of Geology, Quarterly Review, 43* (1830), 411–69; N51 f23; *Table Talk,* June 29, 1833, where STC also ex-

presses belief in both the "rectilinearity" and "undulatory motion" of light; N49 ff29v–30.

35 Royal Society of London, Herschel MS 2 19.

36 *Elements of Chemistry*, trans. R. Kerr (Edinburgh, 1790), pp. xiv, 175–6.

37 D. M. Knight, *Atoms and Elements* (London, 1967). E. C. Patterson, *John Dalton and the Atomic Theory* (New York, 1970), p. 57, tells of a meeting between STC and Dalton. Other accounts of atomism and its reception include D. S. L. Cardwell, ed., *John Dalton and the Progress of Science* (Manchester, 1968); and A. Thackray, *John Dalton: Critical Assessments of His Life and Science* (Cambridge, Mass., 1972).

38 Feren Szabadváry, *History of Analytical Chemistry* (Oxford, 1966); T. H. Levere, *Affinity and Matter: Elements of Chemical Philosophy, 1800–1865* (Oxford, 1971), chap. 6.

39 Knight, *Atoms and Elements*; T. Beddoes, "Letter on certain points of history relative to the component parts of the alkalis," Nicholson's *Journal*, *21* (1808), 139–41.

40 BM Add MS 34225 f148v.

41 N26 ff11v–12 (ca. 1826). There was some speculation that ammonium might be an element like sodium and potassium: ammonium and potassium salts were often similar in many ways. The nature of nitrogen was also problematic. Levere, *Affinity and Matter*, pp. 45–6. STC may be referring to "reversing . . . the *proportions*" of ethyl chloride and water, which would yield ethyl alcohol – "a mad'ning Spirit" – and hydrochloric acid – "a corrosive Acid."

42 N27 f96v (1819); N56 ff15–15v; *Philosophical Lectures*, p. 121. See also *TL*, p. 51 n.

43 CM, Behmen [Boehme], *Works*, *1*, pt. 1, p. 42. Davy's atomism was highly qualified, as explained in Knight, *Atoms and Elements*, and Levere, *Affinity and Matter*, chap. 2. Brande's atomism is explicit in *Manual of Chemistry*, p. 2, and is attacked by STC in N28 f6v.

44 *CL*, 4, 743–4, 760. H. C. Oersted's scheme is in his *Ansicht der chemischen Naturgesetze* (Berlin, 1812). STC's annotated copy is in BM. See also N26 ff86v–7 (ca. 1826).

45 N46 f34.

46 R. A. Foakes, ed., *Coleridge on Shakespeare* (Charlottesville, Va., 1971), p. 65; CM, Berkeley, *Siris* . . . (Dublin and London, 1744), front flyleaf, copy in Beineke Library, Yale University; *CN*, *3*, 4222 (1814).

47 *Friend*, *1*, 470–1.

48 N23 ff31–31v.

49 *CN*, *3*, 4226 (ca. 1815); "On the polar logic," Cornell University Library, Wordsworth Collection.

50 N27 f57v; CM, Steffens, *Geognostisch–geologische Aufsätze*, p. 243, copy in BM; CM, *Beyträge*, p. 262, copy in BM; CM, G. A. Goldfuss, *Handbuch der Zoologie*, 2 vols. (Nürnberg, 1820), *1*, –3– – 1, copy in VCL; CM, F. F. Runge, *Neueste Phytochemische Entdeckungen zur Begründung einer wissenschaftlichen Phytochemie*, 2 pts. (Berlin, 1820–1), pp. 116–17, copy in BM.

51 CM, Oken, *Naturgeschichte*, *1*, 2–4, copy in BM. STC here treats "Diamond or Quartz" as essentially the same, as did Werner and Steffens, for

whom diamond was the first member of the siliceous series of rocks. See Steffens, *Oryktognosie, 1*, xix ff., 1 ff.; and A. G. Werner, "Kurze Klassifikation und Beschreibung der verschiedenen Gebirgsarten," *Abhandlungen der Böhmischen Gesellschaft der Wissenschaften, 2* (1786), 272–97, trans. by A. Ospovat as *Short Classification and Description of the Various Rocks* (New York, 1971). See also N3 ½ ff138–9.

52 *Beyträge*, p. 262.

53 *Table Talk*, June 29, 1833; *TL*, p. 69; N29 f52.

54 *CN*, *3*, 4420. See also N34 f4ᵛ; and N27 ff49–50ᵛ, 54ᵛ.

55 Steffens, *Beyträge*, p. 117; *TL*, p. 91. Cf. BM Egerton MS 2801 f62ᵛ.

56 Steffens, *Beyträge*, p. 138; M. E. Weeks, *Discovery of the Elements*, 6th ed. (Easton, Pa., 1956), pp. 407 ff.; N27 f55; CM, Steffens, *Beyträge*, p. +9.

57 *CN*, *3*, 4420.

58 Weeks, *Discovery*, chap. 27; Davy, "Researches on the oxymuriatic acid, its nature and combinations . . .," *Philosophical Transactions*, *100* (1810), 231–57, *104* (1814), 74–93. STC commented on Davy's work on chlorine when it first appeared: *CL*, *6*, 1027 (Sept. 27, 1811).

59 CM, Steffens, *Beyträge*, back flyleaf, about p. 138; *CL*, *5*, 130.

60 CM, Steffens, *Geognostisch–geologische Aufsätze*, p. +1, about p. 243.

61 N28 f23ᵛ, after Brande, *Manual of Chemistry*, pp. 83, 173.

62 N27 f61 (1819); *CN*, *3*, 4196 [1814].

63 Davy, "Some further observations on a new detonating substance," *Philosophical Transactions*, *103* (1813), 242–51; N27 f49; Royal Institution MS 13j; Levere, *Affinity and Matter*, pp. 45–6.

64 *CN*, *3*, 4196.

65 CM, Thomas Browne, *Pseudodoxia Epidemica* . . ., 3d ed. (London, 1658), pt. 1, p. 37, copy in Berg Collection, New York Public Library.

66 Joseph Priestley, "Experiments on the production of air by the freezing of water," Nicholson's *Journal*, *4* (1801), 193–6; CM, Behmen [Boehme], *Works*, *1*, pt. 2, pp. 36–7 (cf. N27 ff73ᵛ–72ᵛ); CM, Behmen [Boehme], *Works*, *1*, pt. 1, pp. 78–9.

67 M. J. Petry, *Hegel's Philosophy of Nature*, 3 vols. (London and New York, 1970), *2*, 268–70; J. A. Deluc, *Recherches sur les modifications de l'atmosphère*, 2 vols. (Geneva, 1772); Deluc, *Idées sur la météorologie*, 2 vols. (London, 1786–7); Martinus van Marum, *Verhandelingen uitgegeeven door Teyler's Tweede Genootshcap*, vol. *4* (Haarlem, 1787), p. 144. Cf. N23 ff29–29ᵛ.

68 CM, Behmen [Boehme], *Works*, *1*, pt. 1, p. 79. Cf. N27 f54ᵛ.

69 B. Franklin, *New Experiments and Observations on Electricity* (London, 1751); Petry, *Hegel's Philosophy of Nature*, *2*, 266.

70 BM Egerton MS 2800 ff167 (Aug. 1817), 88 (notes to J. Franklin's *Narrative* . . . [London, 1823]).

71 N27 ff62ᵛ–3ᵛ.

72 Thomson, *System of Chemistry*, 6th ed. (Edinburgh, 1820), *2*, 355. Presumably, Brande used the 5th ed., 1817, which I have not seen. The 6th ed. was reviewed in the *Quarterly Journal of Science, Literature, and the Arts*, *11* (1821), 119–71, at 121–2.

73 N27 f67.

74 *Table Talk*, Mar. 18, 1832; *CL*, *5*, 309 (Nov. 1823); Levere, *Affinity and Matter*, chap. 2.

75 J.-C. Guédon, "The still life of a transition: chemistry in the Encyclopédie," Ph.D. diss., University of Wisconsin, 1974.
76 H. Guerlac, "Chemistry as a branch of physics: Laplace's collaboration with Lavoisier," *Historical Studies in the Physical Sciences*, *7* (1976) 193–276; Lavoisier, *Elements of Chemistry*, p. xxi; Berthollet, *Researches into the Laws of Chemical Affinity*, trans. M. Farrell (London, 1804), and *Essai de statique chimique*, 2 vols. (Paris, 1803); Senebier, *Physiologie végétale* . . ., 5 vols. (Geneva, [1800]), *2*, 303; M. Sadoun-Goupil, *Le chimiste Claude-Louis Berthollet, 1748–1822: sa vie, son oeuvre* (Paris, 1977).
77 J. Abernethy, *An Enquiry into the Probability and Rationality of Mr. Hunter's Theory of Life* . . . (London, 1814), pp. 48–51.
78 *Grundzüge*, pp. 48–58; *Geognostisch–geologische Aufsätze*, p. 243; *Beyträge*, pp. 1, 8, 117. Cf. N27, N28, passim.
79 *CN*, *3*, 4244, 4454; *CM*, Nicholson's *Journal*, *26* (1810), 131, copy in VCL MS BT 27.7.
80 *TL*, pp. 59, 91, 56; N25 f85; VCL MS BT 37.7; Levere, *Affinity and Matter*, chap. 2.
81 STC, "On the Divine Ideas," in the Commonplace Book, Huntington Library MS HM 8195 f165; *Friend*, *1*, 94 n. Cf. N26 f16v [1826].
82 N26 f96v; BM Egerton MS 2800 f79v; *Opus Maximum*, VCL S MS 29, *1*, f33.
83 *Opus Maximum*, VCL S MS 29, *1*, ff32, 28; *CN*, *3*, 4244; VCL S MS 5.
84 Allen Debus, *The Chemical Philosophy: Paracelsian Science and Medicine in the Sixteenth and Seventeenth Centuries*, 2 vols. (New York, 1977).
85 M. Crosland, "Lavoisier's theory of acidity," *Isis*, *64* (1973), 306–25; H, Le Grand, "Lavoisier's oxygen theory of acidity," *Annals of Science*, *29* (1972), 1–18.
86 Beddoes had drawn attention to Mayow's work in *Chemical Experiments and Opinions Extracted from a Work Published in the Last Century* (Oxford, 1970). STC's criticism of Lavoisier is in N34 f4.
87 N27 f50v. STC's symbols are explained in N27 ff48v–51v, and in CM, Emanuel Swedenborg, *De Equo Albo de quo in Apocalypsi, cap: xix* . . . (London, 1758), copy in BM.
88 Davy, "Researches on the oxymuriatic acid"; *CN*, *3*, 4171; N29 f127v [1820].
89 Gay-Lussac, "Mémoire sur l'iode," *Annales de chimie*, *91* (1814), 5–160, at 130 ff.; Davy, "On the analogies between the undecompounded substances," *Journal of Science and the Arts*, *1* (1816), 283–8; Davy, *Works*, *5*, 456; M. Crosland, *Gay-Lussac, Scientist and Bourgeois* (Cambridge, 1978); T. H. Levere, "Gay-Lussac and the problem of chemical qualities," in *Actes du colloque Gay-Lussac: la carrière et l'oeuvre d'un chimiste français durant la première moitié du XIXe siècle* (Palaiseau, France 1980), pp. 133–48.
90 *CL*, *2*, 727; *CN*, *2*, 3192 (1807). Cf. CM, Behmen [Boehme], *Works*, *1*, pt. 1, pp. 41–2.
91 N27 ff55v, 59v.
92 Steffens, "Ueber den Oxydations- und Desoxydations-Process der Erde," *Zeitschrift für spekulative Physik*, *1* (1800), 137–68; Oersted, *Ansicht*; CM, ibid.; Levere, *Affinity and Matter*, pp. 131–9; N23 ff28–32v; *CL*, *4*, 772.
93 N28 f2v (1820).

94 Berman, *Social Change*, pp. 131, 132, 136. See also Aubrey A. Tulley, "The chemical studies of William Thomas Brande, 1788–1866," M.Sc. thesis, University of London, 1971; and C. H. Spiers, "William Thomas Brande, leather expert," *Annals of Science*, *25* (1969), 179–201.

95 *CN*, *3*, 4420 (1818); J. Goodfield, *The Growth of Scientific Physiology: Physiological Method and the Mechanist-Vitalist Controversy, Illustrated by the Problems of Respiration and Animal Heat* (London, 1960).

96 "Monologues by the late Samuel Taylor Coleridge (no. I): on life," *Fraser's Magazine*, *12* (1835), 493–6; notebook entitled "Marginalia Intentionalia," ff19–20, in Berg Collection, New York Public Library.

97 X. Bichat, *Physiological Researches on Life and Death*, trans. F. Gold (London, n.d.), pp. 81–3; *TL*, passim; "Marginalia Intentionalia."

98 Senebier, *Physiologie végétale*, *2*, vi.

99 Henry Watts, *A Dictionary of Chemistry* (1859–68; supps. 1872, 1875, 1881), cited in *OED*. Watts's definition, although later than STC's, conforms to STC's usage.

100 Brande, *Manual of Chemistry*, pp. 439–40; N27 f70; Steffens, *Beyträge*, p. 38. For STC's nonchemical use of "educt," see *Lay Sermons*, p. 29 and n.

101 *Botanic Garden*, pt. 1, p. 106 n, referring to Lavoisier, *Traité élémentaire de chimie*, 2 vols. (Paris, 1789), *1*, 132.

102 K. Coburn, ed., *Inquiring Spirit: A New Presentation of Coleridge* (London, 1951), p. 223.

103 H. Davy, "An essay on the generation of phosoxygen, or oxygen gas: and on the causes of the colours of organic beings," in *Contributions to Physical and Medical Knowledge, Principally from the West of England*, ed. T. Beddoes (Bristol, 1799), pp. 151–98, 199–205; Daniel Ellis, *Farther Inquiries into the Changes Induced on Atmospheric Air, by the Germination of Seeds, the Vegetation of Plants, and the Respiration of Animals* (Edinburgh, 1811), pp. 59, 109–219. STC knew Ellis's work; see, e.g., *CM*, Steffens, *Beyträge*, pp. 35–6; N28 f7; *Friend*, *1*, 469 (the reference is not to John Ellis); and BM Add MS 34225 f163^r.

104 *Lay Sermons*, p. 72. This passage is discussed by M. H. Abrams in "Coleridge and the romantic vision of the world," in *Coleridge's Variety: Bicentenary Studies*, ed. J. Beer (London, 1974), pp. 101–33, at 128–30.

105 J. Priestley, *Experiments and Observations on Different Kinds of Air*, 3 vols. (London, 1774–7); "Observations on different kinds of air," *Philosophical Transactions*, *62* (1772), 147–252; Ingenhousz, *Experiments upon Vegetables* (London, 1779); *Philosophical Transactions*, *72* (1782), 438, and *On the Nutrition of Plants . . .* (London, 1796); Senebier, *Physiologie végétale*; de Saussure, *Recherches chimiques sur la végétation* (Paris, 1804), pp. 58, 132–3; Davy, *Works*, *7*, 356; Everett Mendelsohn, *Heat and Life: The Development of the Theory of Animal Heat* (Cambridge, Mass., 1964).

106 Davy, *Agricultural Chemistry*, lecture 3; *Friend*, *1*, 466–70. Knight's major papers are in *Philosophical Transactions*, 1801–8; they were much used by Davy in his *Agricultural Chemistry*.

107 Daniel Ellis, *An Inquiry into the Changes Induced in Atmospheric Air, by the Germination of Seeds, the Vegetation of Plants, and the Respiration of Animals* (Edinburgh, 1807), pp. xii, 46–8; *Edinburgh Review*, *22* (1814), 251–81.

108 T. H. Levere, "Martinus van Marum and the introduction of Lavoisier's

chemistry into the Netherlands," in *Martinus van Marum: Life and Work*, ed. R. J. Forbes (Haarlem, 1969), *1*, 158–286, at 236; de Saussure, *Recherches*, pp. 261 n, 270; Brande, *Manual of Chemistry*, p. 346.

109 N27 ff62–62v. Problems of agricultural science in the generation after STC's were severe enough, as shown by Margaret W. Rossiter, *The Emergence of Agricultural Science: Justus Liebig and the Americans, 1840–1880* (New Haven, 1975).

110 *Agricultural Chemistry*, lecture 7.

111 Davy, *Works*, *7*, 223; Steffens, *Beyträge*, pp. 35–6. Petry, *Hegel's Philosophy of Nature*, *3*, 290, gives an account of Schrader's views.

112 Davy, *Works*, *8*, 38–9; CM, ca. 1823, Steffens, *Beyträge*, pp. 35–6.

113 Brewster, "On the optical and physical properties of Tabasheer," *Philosophical Transactions*, *109* (1819), 283–99; N28 f33.

114 De Saussure, *Recherches*, p. 270; Ellis, *An Inquiry*, pp. 7, 52; Brande, *Manual of Chemistry*, pp. 464–7; N28 ff7–7v. STC conflates Brande's discussion of animal and vegetable respiration.

115 BM Egerton MS 2800 f155v; N27 f63v; Brande, *Manual of Chemistry*, p. 364; N27 f65. N17 f128v and NF° f4v were prompted by a reading of Mirbel on "Cryptogamous and agamous vegetation," *Quarterly Journal of Science, Literature, and the Arts*, *6* (1819), 20–31, 210–26, at 223. NF° f4v also refers to L. N. Vauquelin, "Experiments on mushrooms," *Philosophical Magazine*, *43* (1814), 292–9. N27 f64 was prompted by Brande, *Manual of Chemistry*, p. 368, on fungin. Linnaean botany held on longer in Britain than in France or Germany, where A. L. de Jussieu's system (expounded in his *Genera Plantarum*) was adopted. F. Delaporte, "Des organismes problématiques," *Dix-huitième Siècle*, *9* (1977), 49–59, discusses eighteenth-century debates about "animality" and "vegetality."

116 CM, Oersted, *Ansicht*, p. 70.

117 Brande, *Manual of Chemistry*, pp. 461–2. BM Add MS 34225 f163r also deals with the chemistry of digestion.

118 N28 ff4–9v; Home, "Observations . . .," *Philosophical Transactions*, *97* (1807), 139–79, and "Hints . . .," *Philosophical Transactions*, *99* (1809), 385–91.

119 Brande, *Manual of Chemistry*, pp. 464, 434; C. Hatchett, "Chemical experiments on zoophytes . . .," *Philosophical Transactions*, *90* (1800), 327–402, at 401; N27 f69. STC may also have read Brande's papers on these topics in *Philosophical Transactions*, *99* (1809), 373–84, *101* (1811), 90–114, *102* (1812), 90–114.

120 Brande, *Manual of Chemistry*, 457–60, based on C. Hatchett, "Experiments and observations on shell and bone," *Philosophical Transactions*, *89* (1799), 315–34; N27 ff71–71v, 99–99v.

Chapter 7
Life

1 J. Harris, "Coleridge's readings in medicine," *Wordsworth Circle*, *3* (1972), 85–95; R. Guest-Gornall, "Samuel Taylor Coleridge and the doctors," *Medical History*, *17* (1973), 327–42, at 328. There is a need for a major study of STC and the doctors. See also *CL*, *1*, 320.

2 *CL*, *1*, 518; *CN*, *1*, 1657, 388; H. Davy to Henry Penneck, Jan. 26, 1799,

MS in American Philosophical Society, Philadelphia; Guest-Gornall, "Coleridge and the doctors," p. 330; *CL, 2,* 851–2, 937, *6,* 1025–7; *CN, 3,* 3744.

3 *CL, 4,* 766; N29 f6; James Gillman, *A Dissertation on the Bite of a Rabid Animal: Being the Substance of an Essay Which Received a Prize from the Royal College of Surgeons in London in the Year 1811* (London, 1812).

4 CM, *Edinburgh Medical and Surgical Journal, 14* (1818), *24* (1825), *33* (1830), VCL MS BT 21.12, transcribed by G. Grove; CM, *Quarterly Journal of Foreign and British Medicine and Surgery, 1* (1818–19), VCL MS BT 37, transcribed by Sara Coleridge. N29 f24 discusses A. F. Marcus, "Versuch eines Theorie der Entzündung," *Jahrbücher der Medicin als Wissenschaft, 3* (1808), 51. Other medical topics are addressed in N29 f93; VCL S LT F13.1B; NF° f79v; *CL, 5,* 191; and N26 f20v.

5 A. Haller, "A dissertation on the sensible and irritable parts of animals," *Bulletin of the History of Medicine, 4* (1936), 656–99, reprint of London, 1755, translation of *De partibus corporis humani sensibilius et irritabilius* (Goettingen, 1752). As late as 1828 (N37 f3v), STC was reminding himself to read Haller on irritability.

6 John Neubauer, "Dr. John Brown (1735–88) and early German romanticism," *Journal of the History of Ideas, 28* (1967), 367–82; William Cullen Brown, ed., *The Works of Dr. John Brown* (London, 1804); John Brown, *The Elements of Medicine . . . Translated by the Author . . . with a Biographical Preface by Thomas Beddoes,* 2 vols. (London, 1795); B. Hirschel, *Geschichte der Brown'schen Systems und der Erregungstheorie* (Dresden and Leipzig, 1846); W. R. Trotter, "John Brown and the nonspecific component of human sickness," *Perspectives in Biology and Medicine, 21* (1978), 258–64.

7 *CN, 1,* 388; M. J. Petry, *Hegel's Philosophy of Nature,* 3 vols. (London and New York, 1970), *3,* 379.

8 Neubauer, "Dr. John Brown," pp. 372–3; Schelling, *Werke,* 8 vols. (Munich, 1927–56), *2,* 73; G. B. Risse, "Schelling, 'Naturphilosophie,' and John Brown's system of medicine," *Bulletin of the History of Medicine, 50* (1976), 213–25; J. F. Blumenbach, *Ueber den Bildungstrieb und das Zeugungsgeschäft* (Göttingen, 1781; 3d ed., 1791), *An Essay on Generation,* trans. A. Crichton (London, [1792]), and *The Institutes of Physiology, translated from the Latin [3d ed.] . . .* (London, 1815), p. 38.

9 BM Egerton MS 2801 f120; *CN, 3,* 4226 (1815).

10 N29 f28 (?1824); N59 ff11–14.

11 Those exceptions fascinated STC; see, e.g., N28 f35v. Background is given in R. M. Maniquis, "The puzzling *Mimosa*: sensitivity and plant symbols in romanticism," *Studies in Romanticism, 8* (1969), 129–55.

12 N29 f62v; *Table Talk,* May 2, 1820; N23 f42; N28 f54.

13 Steffens, *Beyträge*; Oken, *Lehrbuch der Naturphilosophie* (Jena, 1809), and *Lehrbuch der Naturgeschichte,* 6 vols. (Leipzig and Jena, 1813; Jena, 1816–26).

14 Oken, *Elements of Physiophilosophy,* trans. A. Tulk (London, 1847), p. 494.

15 *CN, 3,* 4226 (1815); NF° f49; N27 ff37v–42.

16 N21½ f59v; N26 ff19–20v.

17 CM, Behmen [Boehme], *Works, 1,* pt. 1, pp. 120, 49–50, 249–51, copy in BM; CM, H. Boerhaave, *A New Method of Chemistry,* trans. P. Shaw and E. Chambers, 2 pts. in 1 vol. (London, 1727), pt. 1, p. 72, transcribed in

VCL BT 37.3; John Hunter, *A Treatise on the Blood, Inflammation and Gun-shot Wounds, . . . to Which Is Prefixed a Short Account of the Author's Life, by His Brother-in-law Everard Home* (London, 1794), pp. 78–9; Blumenbach, *Institutes of Physiology*, pp. 3–14. William Lawrence, "Life," in Rees's *Cyclopaedia, 20* (1819), gives a good survey of contemporary views on physiology.

18 Cuvier's *Lectures* are referred to in *CN, 3*, 4356–7 and nn (1817).

19 *Von der Weltseele*, p. ix.

20 *Beyträge*, p. 275.

21 Kant, *The Critique of Judgement*, ed. and trans. J. C. Meredith (Oxford, 1973), pt. 2, pp. 21, 22, 24, 39, 74, 82, 87; *Friend, 1*, 497.

22 Kant, *Critique of Judgement*, pt. 2, p. 86; Blumenbach, *A Manual of the Elements of Natural History* (London, 1825), p. 8; Cuvier, *Lectures on Comparative Anatomy, 1*, pp. 6, xx–xxi. Relations between Kant and Blumenbach, and the significance of the biological theories that followed their work, are being developed by T. Lenoir in an important series of papers, of which the first ("Kant, Blumenbach, and Vital Materialism in German biology," *Isis, 71* [1980], 77–108) was published as this book went to press.

23 Hunter, *A Treatise on the Blood*, p. 78; R. Saumarez, *The Principles of Physiological and Physical Science . . .* (London, 1812), p. 42; *BL, 1*, 103 n (chap. 9); J. Abernethy, *Physiological Lectures, Exhibiting a General View of Mr. Hunter's Physiology* (London, 1817), p. 38, *Introductory Lectures Exhibiting Some of Mr. Hunter's Opinions Respecting Life and Diseases* (London, 1815), pp. 16–17. STC criticizes Abernethy's inconsistency in N27 f94.

24 W. Lawrence, *Lectures on Physiology, Zoology and the Natural History of Man . . .* (London, 1819), pp. 7–8.

25 *CL, 4*, 690; *Opus Maximum*, VCL S MS 29, *1*, f9.

26 *Philosophical Lectures*, p. 353; *Letters, Conversations and Recollections of S. T. Coleridge*, ed. T. Allsop, 2 vols. (London, 1836), *1*, 233; BM Egerton MS 2801 f62[v].

27 VCL S MS 13; N26 f19[v]. Cf. CM, Swedenborg, *Prodromus philosophiae ratiocinantis de infinito, et causa finali creationis: deque mechanismo operationis animae et corporis* (Dresden and Leipzig, 1734), 2d blank sheet verso, copy in Huntington Library.

28 *CL, 1*, 294–5.

29 *CN, 2*, 2330 and n.

30 N28 ff44–5.

31 N26 ff19–20[v].

32 N28 ff72–72[v], 1[v].

33 VCL S MS F2.9. Cf. *Fraser's Magazine, 12* (1835), 493–6; and N25 ff86[v]–8.

34 Preformation is discussed in P. J. Bowler, "The impact of theories of generation upon the concept of a biological species in the last half of the eighteenth century," Ph.D. diss., University of Toronto, 1971.

35 Blumenbach, *Natural History*, pp. 8–10, *An Essay on Generation, The Institutes of Physiology*, p. 16, and *Ueber die natürlichen Vershiedenheiten im Menschengeschlechte . . .*, trans. J. G. Gruber (Leipzig, 1798), pp. 69–73; *CN, 3*, 3744.

36 Hunter, *A Treatise on the Blood*, pp. 78–9; *Friend, 1*, 493 n–4 n; N29 f94; N26 f44. Not everyone took this view of Hunter. J. C. Prichard declared

that Hunter's "hypothesis of a vital principle . . . has been proved, by careful examination, to be wanting in every characteristic of a legitimate theory." Prichard, *A Review of the Doctrine of a Vital Principle* . . . (London, 1829), p. 132.

37 STC knew Home's works. See N28 ff3^{v}–4; and *Friend, 1*, 474.

38 Blumenbach, *Short System of Comparative Anatomy*, trans. William Lawrence (London, 1807), p. xv, and *Natural History*, p. 5; CM, Blumenbach, *Ueber die natürlichen Vershiedenheiten*, front flyleaf verso, entry dated Jan. 1828. Cf. N28 ff72^{v}–3 (1820).

39 Cuvier, *Le Règne animal distribué d'après son organisation* . . ., 4 vols. (Paris, 1817), *1*, v–xviii; M. P. Winsor, *Starfish, Jellyfish, and the Order of Life* (New Haven and London, 1976), p. 7.

40 *Lectures on Comparative Anatomy, 1*, xxxvi, 59–60.

41 Oken, *Naturgeschichte, 5*, 1 ff., and *Naturphilosophie*; R. Owen, "Oken," *Encyclopaedia Britannica*, 8th ed, *16*, 498–503; Winsor, *Starfish*, p. 20.

42 N29 ff81–2. STC read, or at least handled, Johann Friedrich Meckel's *System der vergleichenden Anatomie*, 6 vols. in 7 (Halle, 1821–33), CM in New York Public Library. Johann B. Spix, zoologist, wrote *Geschichte und Beurtheilung aller Systeme in der Zoologie* (Nürnberg, 1811). Articles touching on physiology sometimes appeared in J. S. C. Schweigger's *Journal für Chemie und Physik*; I have not found a naturalist named "Sweigger." STC annotated G. A. Goldfuss, *Handbuch der Zoologie*, 2 vols. (Nürnberg, 1820), copy in VCL. A major statement of J. B. Lamarck's classification is his *Philosophie zoologique*, 2 vols. (Paris, 1809). Henri de Blainville's classification is in his *De l'organisation des animaux* (Paris, 1822). André Dumeril was the author of *Zoologie analytique* (Paris, 1809). A good summary of the principal early nineteenth-century systems of classification is Louis Agassiz, *Essay on Classification* (1857; reprint ed., ed. E. Lurie, Cambridge, Mass., 1962). STC's knowledge of French and German zoological schemes grew through his intercourse with J. H. Green. In 1824, Green told his class at the Royal College of Surgeons about Lamarck's classification, with its serial gradations among kinds of animals. He modified this to present his own ascending series of animals, claiming that this integration of parts into a system was in accord with Hunter's work (Royal College of Surgeons MS 67.b.11, notes by W. H. Clift and [Richard Owen]). In a lecture of 1827, he again presented an ascending series of animals, this time acknowledging Cuvier. But his classification is markedly different from Lamarck's, Cuvier's, or indeed that of any other zoologist whose works I have been able to examine. A copy of this scheme, in Green's writing, is in the STC collection at VCL, MS LT 50b. The scheme is broadly compatible with STC's in *TL*.

43 N28 f54.

44 N28 ff57, 54.

45 E.g., *CN, 3*, 3956 ff.; N23 f34^{v}–42; *CN, 2*, 2321–36.

46 N23 ff35^{v}–6; *Opus Maximum*, VCL S MS 29, *2*, f185^{v}; *AR*, pp. 211 ff.; N26 f30^{v}; N59 f9.

47 N28 f69^{v}.

48 *Friend, 1*, 498.

49 *CL, 6*, 595.

50 N23 ff38^{v}–40; *Opus Maximum*, VCL S MS 29, *2*, f61.

51 N25 ff87–87v.
52 P. L. Farber, "The type-concept in zoology during the first half of the nineteenth century," *Journal of the History of Biology*, *9* (1976), 93–119; T. Lenoir, "Generational factors in the origin of *Romantische Naturphilosophie*," *Journal of the History of Biology*, *11* (1978), 57–100.
53 N56 ff9v–10.
54 In N65 ff2–2v, the integration of the substance of the *Theory of Life* in *Opus Maximum* is made explicit.
55 N59 f5.
56 *TL*, pp. 36–7.
57 Although "individuation" is widely used in the Scholastics, in Spinoza, and in Schelling, it is likely that STC's immediate source for this usage was Steffens, e.g., in *Grundzüge*, pp. 70 ff.
58 A taking together, a summary.
59 *TL*, pp. 85–6. Cf. Steffens, *Beyträge*, p. 316.
60 *TL*, p. 85, echoes Genesis 1:26 and 2:7. See also *CL*, *6*, 595 (1826).
61 Phrenology and mesmerism were in STC's day two sciences illuminating the relations of body and mind. Mesmerism in particular suggested to him that there was a higher power of life, manifested through the exertion of will and perhaps one with imagination. See T. H. Levere, "S. T. Coleridge and the human sciences: anthropology, phrenology, and mesmerism," in *Science, Pseudoscience, and Society*, ed. M. P. Hanen, M. J. Osler, and R. G. Weyant (Waterloo, Ont., 1980), pp. 171–92.
62 P. 113.
63 Agricultural chemistry, for example, was well regarded, largely through the vogue for Davy's published lectures. But contemporary ignorance about plant physiology meant that nitrogen fixation and the role of trace metals were glossed over. Similar problems arose in the chemistry of digestion and animal respiration. STC identified and stressed these issues.
64 *Works*, *8*, 347.
65 T. H. Levere, "Coleridge, chemistry, and the philosophy of nature," *Studies in Romanticism*, *16* (1977), 349–79.
66 CM, Hegel, *Logik*, *1*, −4−−1, copy in BM.

INDEX

Printed in the United States
By Bookmasters